Ready, S

MCAS
Mathematics Grade 10
2nd Edition

Staff of Research & Education Association

 *Research & Education Association*

The Curriculum Framework Standards in this book were created and implemented by the Massachusetts Board of Education. For further information, visit the Department of Education website at *http://www.doe.mass.edu/mcas/*.

Research & Education Association
61 Ethel Road West
Piscataway, New Jersey 08854
E-mail: info@rea.com

Ready, Set, Go!®
Massachusetts MCAS Mathematics Test, Grade 10

Published 2011

Copyright © 2009 by Research & Education Association, Inc. Prior edition copyright © 2007 by Research & Education Association, Inc. All rights reserved. No part of this book may be reproduced in any form without permission of the publisher.

Printed in the United States of America

Library of Congress Control Number 2008925149

ISBN 13: 978-0-7386-0441-1
ISBN 10: 0-7386-0441-0

REA® and *Ready, Set, Go!*® are trademarks of Research & Education Association, Inc.

Contents

About Research & Education Association

Founded in 1959, Research & Education Association is dedicated to publishing the finest and most effective educational materials—including software, study guides, and test preps—for students in middle school, high school, college, graduate school, and beyond. Today, REA's wide-ranging catalog is a leading resource for teachers, students, and professionals. We invite you to visit us at *www.rea.com* to find out how REA is making the world smarter.

Acknowledgments

We would like to thank REA's Larry B. Kling, Vice President, Editorial, for supervising development; Pam Weston, Vice President, Publishing, for setting the quality standards for production integrity and managing the publication to completion; Michael Reynolds, Managing Editor, for project management and preflight editorial review; Mel Friedman, Math Editor, for editorial contributions; Christine Saul, Senior Graphic Artist, for cover design; and Jeff LoBalbo, Senior Graphic Artist, for post-production file mapping.

We also gratefully acknowledge Marcy Muzykewicz, a math teacher at Charlestown High School, Boston, for content development, and Caragraphics for typesetting.

Introduction

Passing the MCAS Mathematics Test

About This Book

This book will provide you with an accurate and complete representation of the MCAS Mathematics test. In it you will find lessons that are designed to provide you with the information and strategies needed to do well on the test. Two practice tests based on the official MCAS are provided. The practice tests contain every type of question you can expect to encounter on the MCAS. After each test, you will find an answer key with detailed explanations designed to help you completely understand the test material.

About the Test

Who Takes These Tests and What Are They Used For?

The MCAS, or Massachusetts Comprehensive Assessment System, is given in grades 3 through 11. The MCAS is aligned with the Massachusetts Curriculum Framework standards that were adopted by the State Board of Education to test students' knowledge in mathematics, English language arts, science and technology/engineering, and history and social science. Students must demonstrate that they have achieved the standards on the grade 10 MCAS English language arts and math tests in order to earn a Massachusetts high school diploma.

Is There a Registration Fee?

No.

When and Where Is the Test Given?

The MCAS is administered annually. Testing periods are scheduled for the spring of each year, and retests are administered in the fall. All students must pass the tenth grade tests in order to graduate. Students who do not pass these tests will be retested in grade 11 and in grade 12, if necessary. Tests are given in school. The grade 10 MCAS Mathematics test is administered over two consecutive days (in two one-hour sessions on each day).

Test Accommodations and Special Situations

All students educated with public funds are required to take the MCAS. Seldom is an exemption granted. Special education and disabled students are required to take the MCAS. Every effort is made to provide a level playing field for students with disabilities who are taking the MCAS and seeking a standard high school diploma. For example, the timing or scheduling of the test can be altered to accommodate a student's medical needs. Changes to suit a student's specific needs can also be made to the test setting, how the test is presented, and how the student responds to test questions. Students with significant disabilities who are unable to take the standard MCAS tests even with accommodations are offered the MCAS Alternate Assessment (MCAS-Alt). English language learners (ELL) are required to take the MCAS. Grade 10-12 Spanish-speaking ELL students who have been in school in the continental U.S. for less than three years can choose to participate in the MCAS Spanish/English Mathematics Grade 10 Test and Mathematics Retest if they can read and write at or near grade level in Spanish.

Additional Information and Support

Additional resources to help you prepare to take the MCAS can be found on the Massachusetts Department of Education website at *www.doe.mass.edu*.

How to Use This Book

What Do I Study First?

Read over the chapters and the suggestions for test taking. Studying the chapters thoroughly will reinforce the basic skills you need to do well on the test. Be sure to take the practice tests to become familiar with the format and procedures involved with taking the actual MCAS.

When Should I Start Studying?

It is never too early to start studying for the MCAS; the earlier you begin, the more time you will have to sharpen your skills. Do not procrastinate! Cramming is *not* an effective way to study, because it does not allow you the time you need to master the test material. The sooner you learn the format of the exam, the more time you will have to familiarize yourself with the exam content.

Overview of the MCAS in Mathematics

Test takers will encounter 21 to 26 questions in each session of the mathematics portion of the MCAS. Thus, you can expect to see as many as 52 questions on the full test, though only 21 questions from each session will actually be scored; the other questions are being tried out for future tests. You won't know which items will be scored and which are experimental, and there's no reason to worry about the difference; just do your best on all the questions. These questions are based on the learning standards in the Massachusetts Mathematics Curriculum Framework (2000). The Framework identifies the following five major content strands:

- Number Sense and Operations
- Patterns, Relations, and Algebra
- Geometry
- Measurement
- Data Analysis, Statistics, and Probability

Each session includes multiple-choice and open-response questions. Session 1 also includes short-answer questions. Short-answer and open-response items require students to generate a written response. A short-answer item requires a numeric solution to a straightforward problem. Students are provided with a small box on their answer sheet in which the answer must be placed. An open-response item requires students to solve a more complex problem and to provide a more in-depth response. Students are typically asked to show their work or calculations, explain their reasoning, and justify the procedures used.

Open-response items may require 5 to 15 minutes to complete, and responses receive a score of 0, 1, 2, 3, or 4 points.

Open-response questions require students to provide evidence of content knowledge; understanding of mathematical concepts, principles, and procedures; and problem-solving and mathematical communication skills. Answers will be individually read and evaluated against a scoring guide, or "rubric," that is developed for each individual test question. Student answers to open-response questions will be judged on the following, where relevant:

- Accuracy of solutions
- Knowledge of mathematical facts and procedures
- Understanding of mathematical concepts
- Quality of mathematical reasoning
- Efficiency of application of mathematical procedures
- Correct use of mathematical communication terms, diagrams, and symbols

Open-response answers will not be judged directly for writing skills (grammar, punctuation, or other conventions). But it is important for you to realize that your writing ability is *indirectly* important, because it is through your writing alone that you must provide a **clear** and **accurate** response that effectively presents your mathematical reasoning, concepts, and solutions.

Standards

Numbers, Number Sense, and Operations

10.N.1   Identify and use the properties of operations on real numbers, including the associative, commutative, and distributive properties; the existence of the identity and inverse elements for addition and multiplication; the existence of nth roots of positive real numbers for any positive integer n; and the inverse relationship between taking the nth root of and the nth power of a positive real number.

10.N.2   Simplify numerical expressions, including those involving positive integer exponents or the absolute value, e.g., $3(2^4 - 1) = 45$, $4 \cdot |3 - 5| + 6 = 14$; apply such simplifications in the solution of problems.

10.N.3   Find the approximate value for solutions to problems involving square roots and cube roots without the use of a calculator, e.g., $\sqrt{3^2 - 1} \approx 2.8$.

10.N.4   Use estimation to judge the reasonableness of results of computations and of solutions to problems involving real numbers.

Data Analysis, Statistics, and Probability

10.D.1   Use appropriate statistics (e.g., mean, median, range, and mode) to communicate information about the data. Use these notions to compare different sets of data. Select, create, and interpret an appropriate graphical representation (e.g., scatterplot, table, stem-and-leaf plots, box-and-whisker plots, circle graph, line graph, and line plot) for a set of data.

10.D.2   Approximate a line of best fit (trend line) given a set of data (e.g., scatterplot). Use technology when appropriate.

10.D.3   Describe and explain how the relative sizes of a sample and the population affect the validity of predictions from a set of data.

Geometry

10.G.1   Identify figures using properties of sides, angles, and diagonals. Identify the figures' type(s) of symmetry.

10.G.2   Draw congruent and similar figures using a compass, straight edge, protractor, and other tools such as computer software. Make conjectures about methods of construction. Justify the conjectures by logical arguments.

10.G.3   Recognize and solve problems involving angles formed by transversals of coplanar lines. Identify and determine the measure of central and inscribed angles and their associated minor and major arcs. Recognize and solve problems associated with radii, chords, and arcs within or on the same circle.

10.G.4   Apply congruence and similarity correspondences (e.g., $\triangle ABC \cong \triangle XYZ$) and properties of the figures to find missing parts of geometric figures, and provide logical justification.

10.G.5   Solve simple triangle problems using the triangle angle sum property and/or the Pythagorean theorem.

10.G.6   Use the properties of special triangles (e.g., isosceles, equilateral, 30°-60°-90°, 45°-45°-90°) to solve problems.

10.G.7   Using rectangular coordinates, calculate midpoints of segments, slopes of lines and segments, and distances between two points, and apply the results to the solutions of problems.

10.G.8   Find linear equations that represent lines either perpendicular or parallel to a given line and through a point, e.g., by using the "point-slope" form of the equation.

10.G.9   Draw the results, and interpret transformations on figures in the coordinate plane, e.g., translations, reflections, rotations, scale factors, and the results of successive transformations. Apply transformations to the solutions of problems.

10.G.10 Demonstrate the ability to visualize solid objects and recognize their projects and cross sections.

10.G.11 Use vertex-edge graphs to model and solve problems.

Measurement

10.M.1   Calculate perimeter, circumference, and area of common geometric figures such as parallelograms, trapezoids, circles, and triangles.

10.M.2   Given the formula, find the lateral area, surface area, and volume of prisms, pyramids, spheres, cylinders, and cones, e.g., find the volume of a sphere with a specified surface area.

10.M.3   Relate changes in the measurement of one attribute of an object to changes in other attributes, e.g., how changing the radius or height of a cylinder affects its surface area or volume.

10.M.4   Describe the effects of approximate error in measurement and rounding on measures and on computed values from measurements.

Patterns, Relations, and Algebra

10.P.1   Describe, complete, extend, analyze, generalize, and create a wide variety of patterns, including iterative, recursive (e.g., Fibonacci numbers), linear, quadratic, and exponential functional relationships.

10.P.2   Demonstrate an understanding of the relationship between various representations of a line. Determine the line's slope and x- and y-intercepts from its graph or from a linear equation that represents a line. Find a linear equation describing a line from a graph or a geometric description of the line, e.g., by using the "point-slope" or "slope y-intercept" formulas. Explain the significance of a positive, negative, zero, or undefined slope.

10.P.3   Add, subtract, and multiply polynomials. Divide polynomials by monomials.

10.P.4   Demonstrate facility in symbolic manipulation of polynomial and rational expressions by rearranging and collecting terms; factoring (e.g., $a^2 - b^2 = (a + b)(a - b)$, $x^2 + 10x + 21 = (x + 3)(x + 7)$, $5x^4 + 10x^3 - 5x^2 = 5x^2(x^2 + 2x - 1)$; identifying and canceling common factors in rational expressions; and applying the properties of positive integer exponents.

10.P.5   Find solutions to quadratic equations (with real roots) by factoring, completing the square, or using the quadratic formula. Demonstrate an understanding of the equivalence of the methods.

10.P.6 Solve equations and inequalities including those involving absolute value of linear expressions (e.g., $|x - 2| > 5$) and apply to the solution of problems.

10.P.7 Solve everyday problems that can be modeled using linear, reciprocal, quadratic, or exponential functions. Apply appropriate tabular, graphical, or symbolic methods to the solution. Include compound interest, and direct and inverse variation problems. Use technology when appropriate.

10.P.8 Solve everyday problems that can be modeled using systems of linear equations or inequalities. Apply algebraic and graphical methods to the solution. Use technology when appropriate. Include mixture, rate, and work problems.

Test-Taking Strategies

What to Do Before the Test

- **Pay attention in class.**

- **Carefully work through the chapters of this book.** Mark any topics that you find difficult, so that you can focus on them while studying and get extra help if necessary.

- **Take the practice tests and become familiar with the format of the MCAS.** When you are practicing, simulate the conditions under which you will be taking the actual test. Stay calm and pace yourself. After simulating the test only a couple of times, you will feel more confident, and this will boost your chances of doing well.

- **Students who have difficulty concentrating or taking tests in general may have severe test anxiety.** Tell your parents, a teacher, a counselor, the school nurse, or a school psychologist well in advance of the test if this applies to you. They may be able to give you some useful strategies that will help you feel more relaxed and then be able to do your best on the test.

What to Do During the Test

- **Read all of the possible answers.** Even if you think you have found the correct response, do not automatically assume that it is the best answer. Read through each answer choice to be sure that you are not making a mistake by jumping to conclusions.

- **Use the process of elimination.** Go through each answer to a question and eliminate as many of the answer choices as possible. By eliminating two answer choices, you have given yourself a better chance of getting the item correct, because there will only be two choices left from which to make your selection. Sometimes a question will have one or two answer choices that are a little odd. These answers will be obviously wrong for one of several reasons: they may be impossible given the conditions of the problem, they may violate mathematical rules or principles, or they may be illogical.

- **Work on the easier questions first.** If you find yourself working too long on one question, make a mark next to it on your test booklet and continue. After you have answered all of the questions that you know, go back to the ones you have skipped.

- **Be sure that the answer oval you are marking corresponds to the number of the question in the test booklet.** The multiple-choice sections are graded by machine, so marking one wrong answer can throw off your answer key and your score. Be extremely careful.

- **Work from answer choices.** You can use a multiple-choice format to your advantage by working backward from the answer choices to solve a problem. This strategy can be helpful if you can just plug the answers into a given formula or equation. You may be able to make a better choice on a difficult problem if you eliminate choices that you know do not fit into the problem.

- **If you cannot determine what the correct answer is, answer the question anyway.** The MCAS does not subtract points for wrong answers, so be sure to fill in an answer for every question. It works to your advantage because you could guess correctly and increase your score.

The Day of the Test

On the day of the test, you should wake up early (it is hoped after a good night's rest) and have a good breakfast. Make sure to dress comfortably, so that you are not distracted by being too hot or too cold while taking the test. Make sure to give yourself enough time to arrive at your school early. This will allow you to collect your thoughts and relax before the test, and will also spare you the anguish that comes with being late.

Chapter 1

Number Sense and Operations, Part 1

$$3 + 7 < 16$$
$$\sqrt{8^2 + 5}$$
$$20 - 9 \times 6$$

Standards

10.N.1   Identify and use the properties of operations on real numbers, including the associative, commutative, and distributive properties; the existence of the identity and inverse elements for addition and multiplication; the existence of *n*th roots of positive real numbers for any positive integer *n*; and the inverse relationship between taking the nth root of and the nth power of a positive real number.

10.N.3   Find the approximate value for solutions to problems involving square roots and cube roots without the use of a calculator, e.g., $\sqrt{3^2 - 1} \sim 2.8$.

Questions about Number Sense and Operations might ask you about the properties of real numbers. You will need to know, for example, the difference between the associative, commutative, and distributive properties. You will also need to know how to determine the value of a number raised to a power and the square root of a number.

Types of Numbers

For some questions on the MCAS, you need to be able to recognize different kinds of numbers and put them in order by size. Some questions will ask you to choose a number that is the greatest or the least or to choose a number that is equivalent to another number expressed in a different form.

Integers are whole numbers and their opposites. A number's opposite is its negative. The following numbers are pairs of opposites:

1	−1
2	−2
3	−3
4	−4
5	−5

Real numbers are numbers that can be placed on a number line. Real numbers are grouped into two categories: rational and irrational. **Rational numbers** include whole numbers, fractions, and decimals, even if these numbers are repeating, meaning they don't terminate. **Irrational numbers** are decimals that don't repeat or terminate in a logical manner. These numbers are irrational:

2.3459564646332
0.999088432

Absolute Value

The **absolute value** of a number is the number of places it is from zero. To find a number's absolute value, it helps to imagine a number line. Look at the number line below. You can see that the absolute value of −2 is 2 because −2 is 2 units from 0. Some questions on the MCAS may ask you to find the absolute value of a number.

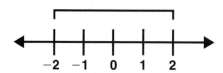

Equivalent Numbers

Numbers that are **equivalent** have the same value. With some numbers, it is easy to see that they are equivalent. For example, you know that $7 = 7$, and you probably know that $\frac{5}{5}$ is equivalent to 1.

Determining whether numbers are equivalent when they are in different forms is more difficult, however. You might not know right away that 8^3 is equivalent to 512.

The best way to determine whether numbers are equivalent is to put them into the same form. The sections that follow present several guidelines to help you do this.

Fractions

A **fraction** represents how many parts you have of an object that contains a number of equal parts. The numerator (top number) of a fraction tells how many parts you have. The denominator (bottom number) tells the number of equal parts that are in the object. For example, the fraction $\frac{2}{3}$ tells you that you have 2 out of 3 parts.

If the denominators of two fractions are the same, the fraction with the *larger* numerator is the larger fraction. For example, $\frac{3}{7}$ is larger than $\frac{2}{7}$.

Equivalent fractions are fractions that have the same value. For example, $\frac{1}{2}$ is equivalent to $\frac{2}{4}$ and $\frac{3}{6}$.

To determine whether fractions are equivalent, multiply the numerator and denominator of one fraction by the same number. The number that you multiply the fraction by should make the numerators of both fractions the same. For example, to determine whether the fractions $\frac{1}{2}$ and $\frac{2}{4}$ are equivalent, multiply $\frac{1}{2}$ by $\frac{2}{2}$.

$$\frac{1}{2} \times \frac{2}{2} = \frac{2}{4}$$

Therefore, $\frac{1}{2}$ and $\frac{2}{4}$ are equivalent. Let's try another: $\frac{2}{3}$ and $\frac{4}{5}$.

$$\frac{2}{3} \times \frac{2}{2} = \frac{4}{6}$$

Therefore, the fractions $\frac{4}{5}$ and $\frac{2}{3}$ are not equivalent.

If two fractions have different numerators and denominators, you can determine whether they are equivalent by making the denominators the same. To do this, you must realize that 1 times any number equals the same number. Multiply one of the fractions by the equivalent of 1 so that the denominators of the two fractions are the same. Then compare the results to see whether the fractions are equivalent. $\frac{2}{3} \times \frac{5}{5} = \frac{10}{15}$, whereas $\frac{3}{5} \times \frac{3}{3} = \frac{9}{15}$. Thus, $\frac{2}{3}$ is larger than $\frac{3}{5}$.

If you're asked to compare two or more mixed numbers (a mixed number has a whole number and a fraction, such as $1\frac{1}{2}$), the one with the larger whole number is the greater number. For example: $2\frac{1}{3}$ is greater than $1\frac{1}{3}$.

If the whole number parts are the same, use the method you just learned to compare the fractional parts to determine which mixed number is larger.

Decimals

A mixed decimal number, such as 3.14, includes a decimal point and has two parts. The part to the left of the decimal point is a whole number, and the part to the right of the decimal point is called a **decimal**. A decimal is not a whole number, but is a portion of a whole number, and has a value less than 1. Therefore, the number 3 is greater than the number .33. The decimal .33 can also be written as 0.33, indicating that there is no whole number part to the decimal.

The decimal system is based on the number 10 (this probably has to do with the fact that most humans have 10 fingers). Each digit in a decimal number has a value assigned to its "place." To the left of the decimal point, the digits appear as you are used to seeing them (ones, tens, hundreds, etc.), but to the right of the decimal point they are fractions, so they are tenths, hundredths, thousandths, etc. (The decimal system is discussed further in the section on scientific notation later in this chapter.)

So for the decimal number 4.25, the 4 is a whole number. However, the 2 is tenths and the 5 is hundredths, or you could put these last two together and say 25 hundredths. You would read 4.25 as "four point two five," or "four and twenty-five hundredths."

You may be asked to compare decimals on the MCAS. Do you know which is greater, .334 or .3? To determine which decimal is greater, align the decimal points vertically, like this:

.334
.3

Then fill in the empty place values with zeros so both numbers have the same number of digits before you do the comparison:

.334
.300

Which decimal is greater? If you said .334, you're correct! This method is similar to comparing whole numbers, but you must remember to add the zeros to the ends of the decimal fractions so each decimal has the same number of digits.

If you're asked to compare two mixed decimals, the decimal with the greater whole number is always larger. For example, 2.334 is greater than 1.945. If you're asked to compare two mixed decimals with the same whole number, use the method you just learned to compare the decimals to determine which is greater.

For example, to compare 1.4 and 1.36, align the decimal points of the numbers vertically and fill in with zeros if necessary:

1.40
1.36

Now you can see that 1.4 is definitely greater.

On the MCAS, you will be asked to compare numbers that are in different forms, such as fractions and decimals. When comparing a fraction and a decimal, convert the fraction into a decimal by dividing the denominator into the numerator. You can use your calculator to do this. Try converting the fractions below into decimals on your calculator.

$$\frac{3}{4} = .75 \qquad \frac{5}{6} = .8333333$$

Terminating decimals are decimals that stop. For example, .75 is a terminating decimal. **Repeating decimals** keep on going. .8333333 is a repeating decimal (the 3 keeps repeating). To indicate that a decimal is a repeating decimal, a line is usually placed over the last number, like this: $.83\overline{3}$.

Percents

A **percent** has a percent sign (%) and refers to how much "out of 100" a number is. For example, 75% means 75 out of 100.

Determining which of two (or more) percents is greater is sometimes easy. For example, 75% is obviously greater than 65%. Most often on the MCAS, however, you will be asked which number in a different form is equivalent to or greater than a percentage. You might be asked, for example, whether 75% is equivalent to $\frac{3}{4}$. (It is!)

As you learned earlier in this chapter, to find an equivalent number or to compare numbers, convert the numbers to the same form. Usually, it is easiest to convert to decimals to compare numbers of different forms.

To convert a percent into a decimal, move the decimal point to the left two places. (These places represent the two zeros in 100.) Look at the following examples:

$$32\% = .32$$
$$75\% = .75$$
$$210\% = 2.10$$

Fill in with zeros if necessary:

$$5\% = .05$$
$$.3\% = .003$$

Now, suppose you need to convert a decimal to a percent. You would move the decimal point two places to the *right*, and add the percentage sign.

$$.20 = 20\%$$

If you need to convert a percent to a fraction, put the percentage over 100. Then **reduce the fraction**, if possible, by the following method: Think of a number that divides evenly into both the numerator and denominator (called a **common factor**). Do that division, and the results for each part give you the reduced fraction.

So for 20%, the calculation would be: $20\% = \frac{20}{100} = \frac{1}{5}$, by dividing both numerator and denominator by 20.

Let's say that you didn't recognize right off that 20 divides into both 20 and 100 in the above example. Let's say you thought of 10 instead. Then the calculation would be: $20\% = \frac{20}{100} = \frac{2}{10}$. Perhaps now you see that 2 will divide into both the 2 and 10. The result will be $\frac{2}{10} = \frac{1}{5}$, the same result as when you reduced the fraction by using 20 as the common factor: $20\% = \frac{20}{100} = \frac{2}{10} = \frac{1}{5}$.

Powers

To determine if numbers are equivalent on the MCAS, you may have to raise a number to a certain power. The **power** (indicated by a raised number, called an **exponent**) tells you how many times the number appears when it is multiplied by itself. Thus, the first power of any number is equal to itself. For example:

$$8^1 = 8$$

When you raise a number to the second power, you **square** the number. When you square a number, you multiply it by itself, as in this example:

$$8^2 = 8 \times 8$$

When you raise a number to the third power, you **cube** the number. To cube a number, multiply it by itself and then by itself again:

$$8^3 = 8 \times 8 \times 8$$

You can keep raising numbers to higher and higher powers, as in this example:

$8^9 = 8 \times 8 \times 8 \times 8 \times 8 \times 8 \times 8 \times 8 \times 8$, where the number 8 appears nine times.

Square Roots and Radicals

The **square root** of a number is the **inverse operation** (opposite) of squaring the numbers (multiplying the number by itself). For example, $\sqrt{144} = 12$.

Not every number is a perfect square. This means you might not always get a whole number when you find the square root. For example:

$$\sqrt{3} \approx 1.73205081$$

Radicals are symbols for roots. When adding or subtracting radicals, keep the number under the radical, called the **radicand**, the same and add or subtract the coefficients, the numbers in front of the square root sign. For example,

$$2\sqrt{3} + 6\sqrt{3} = 8\sqrt{3}$$

When multiplying or dividing radicals, follow the same rules you would when multiplying or dividing whole numbers. For example,

$$\sqrt{3} \times \sqrt{3} = \sqrt{9}$$
$$\sqrt{8} \times \sqrt{2} = \sqrt{16}$$
$$2\sqrt{2} \times 4\sqrt{5} = 8\sqrt{10}$$

Let's Review 1: Equivalent Numbers

Complete each of the following questions about equivalent numbers. Use the Tip below each question to help you choose the correct answer. When you finish, check your answers with those at the end of Chapter 1. Some questions on the actual exam do allow the use of a calculator. Students should try as many questions as possible without a calculator.

1   **Which of the following is closest to the value of $\sqrt{21}$?**

A.   4.3

B.   4.4

C.   4.6

D.   4.8

> The number 21 is not a perfect square. Choose the number that is closest.

2   **Which is another way to express 144?**

A. 12^2

B. 4^4

C. 14.4×10^2

D. 8^3

> If you're not sure of the correct answer, begin by eliminating those that you know are incorrect.

3   **Which fraction is closest to .45?**

A. $\dfrac{1}{45}$

B. $\dfrac{9}{20}$

C. $\dfrac{1}{4}$

D. $\dfrac{1}{2}$

TIP

When you convert a decimal to a fraction, move the decimal point two places to the right and use 100 as the denominator. Then reduce the fraction. You can also solve this problem quickly by using your calculator to find the decimal equivalent of each fraction by dividing the denominators into the numerators and comparing the decimal results.

4   **What is .25 expressed as a percent? Show your work.**

TIP

To convert a decimal to a percent, move the decimal point two places to the right.

5   **The side of a right triangle is $\sqrt{32}$ inches. Which point is closest to $\sqrt{32}$ on the number line?**

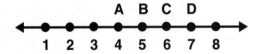

A.   A

B.   B

C.   C

D.   D

TIP

Find the square root of 32. Then choose the point that is closest to this number.

Order of Operations

Some questions on the MCAS will involve more than one operation. To answer these questions correctly, you have to perform operations in the correct order. Look at this problem:

$$2 + 8 \times 3$$

If you solve the problem from left to right, the answer is 30.

However, if you multiply first, $8 \times 3 + 2$, the answer is 26. This example uses the correct order of operations.

Perform operations in this order:

1.   Perform operations in parentheses first.

2.   Perform powers from left to right next.

3.   Then multiply and divide from left to right.

4.   Lastly, add and subtract from left to right.

Properties and Operations of Real Numbers

Numbers have to be added and multiplied in a certain order to correctly solve a problem. On the MCAS, this is referred to as a **property**. Questions about properties on the MCAS will often involve the commutative, associative, and distributive properties of real numbers. You should also know about identity elements, inequalities, and inverse operations.

Commutative Property

The **commutative property** applies to both addition and multiplication. It says that numbers can be added or multiplied in any order.

When the commutative property is applied to addition, for example, $3 + 6$ and $6 + 3$ are the same; they both equal 9.

Sometimes letters, called variables, are used instead of numbers. The commutative property can also be used with letters.

$$x + y = y + x$$

When the commutative property is applied to multiplication you see, for example, that 2×6 and 6×2 are the same; they both equal 12. Using letters, the commutative property can be written two ways:

$$x \times y = y \times x \text{ or } xy = yx$$

Questions on the MCAS might ask you to identify the correct use of the commutative property, or they might ask you to identify a situation that demonstrates the commutative property.

A question on the MCAS about the commutative property might look like the following:

Example

Daryl plans to spend $6.00 per person for a surprise party for his mother. Ten people are coming to the party. His total cost can be expressed as 6.00×10.

Use the commutative property to write an equivalent expression.

A. $6.00 + 10

B. $10 \times $6.00

C. $6.00 \times 10

D. $\dfrac{\$6.00}{10}$

To correctly answer this question, you need to know that the commutative property of multiplication means that two numbers can be multiplied in any order. The only answer that shows a different order for multiplication of the two numbers is choice B.

Associative Property

The **associative property** applies to both addition and multiplication. It says that the numbers being added or multiplied can be combined in any order.

When it is applied to addition, it means that when you add three numbers, you can add them in any grouping. For example, $5 + 2 + 3$ can be solved like this:

$$(5 + 2) + 3 \text{ or } 5 + (2 + 3)$$

where the parentheses indicate which addition to do first. Both ways add up to 10.

Using letters, we can write the associative property for addition this way:

$$a + b + c = (a + b) + c = a + (b + c).$$

The associative property is true for multiplication also. For example, $2 \times 3 \times 4$ can be solved like this:

$$(2 \times 3) \times 4 \text{ or } 2 \times (3 \times 4)$$

where the parentheses indicate which multiplication to do first. Both ways give the same answer, 24.

Using letters, the associative property for multiplication can be written in two ways:

$$(a \times b) \times c = a \times (b \times c) \text{ or } (ab)c = a(bc)$$

Example

During her first year on the basketball team, Sarah scored x points. During her second year, she scored 65 points, and during her third year, she scored 75 points. Her total points for the three years could be expressed as $x + (65 + 75)$.

Use the associative property to write an equivalent expression.

A. $x = 65 + 75$

B. $65x + 75x$

C. $x(65 + 75)$

D. $(x + 65) + 75$

This question asks you about the associative property of addition. Multiplication should not be involved, so you can eliminate answer choices B and C. Answer choice A does not add the three values. Answer choice D is correct.

Distributive Property

The **distributive property** is used when an addition or a subtraction problem is being multiplied by a number. According to the distributive property, you can either (a) add or subtract first and then multiply or (b) multiply first and then add or subtract. Either way, the answer will be the same. Look at the following:

$$(2 + 3) \times 5 = 5 \times 5 = 25 \text{ or}$$
$$2 \times 5 = 10, 3 \times 5 = 15, 10 + 15 = 25$$

Identity Elements

Identity elements leave other numbers unchanged when they are combined with them. Think about addition. What can you add to any number and still get that same number? Zero! Zero is the identity element for addition.

The number 1 is the identity element for multiplication. You can multiply any number by 1 and still get that same number. You might see a question on the MCAS asking you to choose the identity element for either addition or multiplication.

Example

Celia knows that the identity element for addition is 0. What is the identity element for multiplication?

A. 0

B. 1

C. $\dfrac{1}{x}$

D. $\dfrac{0}{x}$

The identity element for multiplication is always 1, so answer choice B is correct.

Other Properties

You might also see a question on the MCAS about the inverse property and the property of equality or inequality. The **inverse** of something is simply the opposite. The additive inverse of 52 is -52. If you add a number and its inverse together, you get 0, the identity element for addition.

For multiplication, the inverse is a little bit different. If you multiply a number and its inverse, you should get 1, the identity element for multiplication. So, the multiplicative inverse of 7 is $\dfrac{1}{7}$ because if you multiply $\dfrac{7}{1}$ by $\dfrac{1}{7}$, you get 1. Note that the multiplicative inverse of a number is also called its **reciprocal**.

Example

Which of the following numbers illustrates the inverse property of addition?

A. $3 + -3 = 1$

B. $3 + -3 = 0$

C. $3 \times -3 = 1$

D. $3 \times -3 = 0$

The correct answer is B because it involves the identity element for addition, 0. Choices C and D can be eliminated right away because they don't involve addition.

When two or more things are equal, as in an equation, an equal sign is used. However, as you know, not all relationships are equal. For example, $7 + 1 \neq 8 + 1$, where $\neq$ means "does not equal," usually indicated by an **inequality** sign. The following are inequality signs you should know:

$>$	greater than
$<$	less than
$\geq$	greater than or equal to
$\leq$	less than or equal to

Example

Which of the following is a correct statement?

A. $7 > 8$

B. $3 + 4 > 8$

C. $3 + 4 < 5$

D. $3 + 4 < 8$

The correct answer is D: $3 + 4$, or 7, is less than 8. All of the other choices are false.

Let's Review 2: Equivalent Expressions

Complete each of the following questions about equivalent expressions. Use the Tip below each question to help you choose the correct answer. When you finish, check your answers with those at the end of the chapter.

1 During one year, a school enrolled a number of new students, expressed as *n*. During the next year, the school enrolled 38 new students. During the following year, it enrolled 92 more students. The principal wrote this expression to show the number of new students enrolled at the school over three years: $n + (38 + 92)$. According to the associative property, which of the following is an equivalent expression?

A. $n(38 + 92)$

B. $(n + 38) - 92$

C. $(n + 38) + 92$

D. $38 + 92n$

TIP

Remember that the associative property of addition says that you can group numbers in any order.

2 Which of the following numbers is the additive inverse of -24?

A. -24

B. $\dfrac{1}{24}$

C. 24

D. 1

TIP

Choose the number that is the opposite of -24.

3 Which equation below illustrates the commutative property of multiplication?

A. $xz = zx$

B. $(xy)z = x(yz)$

C. $\dfrac{xy}{z} = \dfrac{x}{yz}$

D. $(x + y)z = x(y + z)$

TIP

According to the commutative property of multiplication, you can multiply numbers or letters in any order.

4 Which of the following numbers illustrates the inverse property of multiplication?

A. $-5 + 5 = 0$

B. $5 \times \dfrac{1}{5} = 0$

C. $5 \times \dfrac{1}{5} = 1$

D. $-5 \times 5 = 1$

TIP

Remember that for multiplication, a number multiplied by its inverse should equal 1.

5 Which property of real numbers is demonstrated by the equation below?

$$a(x + y) = ax + ay$$

A. associative property of addition

B. commutative property of addition

C. inverse property of addition

D. distributive property

TIP

Try to remember the property that applies to an addition or multiplication problem.

Chapter 1 Practice Problems

Complete each of the following practice problems. Check your answers at the end of this chapter. Be sure to read the answer explanations!

1   **Which value is the greatest?**

A. 7^3

B. 7^4

C. 8^6

D. 9^5

2   **Which answer is .64 written as a percent?**

A.   .64%

B.   6.4%

C.   64%

D.   640%

3   **Which of the following numbers, when multiplied by 4^3, is equal to 1?**

A. 4^{-3}

B. -4^{-3}

C. 4^3

D. -4^3

4 Which of the following properties of real numbers is demonstrated by the equation below?

$$a + b + c = (a + b) + c = a + (b + c)$$

A. associative property of addition

B. commutative property of addition

C. inverse property of addition

D. distributive property

5 Which of the numbers below is the greatest?

A. $\sqrt{121}$

B. $\dfrac{121}{12}$

C. 1.21×10^3

D. 12% of 121

6 Which of the following is an example of the identity property of multiplication?

A. $6 \times 0 = 0$

B. $6 \times \dfrac{1}{6} = 1$

C. $6 \times 1 = 6$

D. $6 + 0 = 6$

7 Which value is the greatest?

A. 3^4

B. 2^5

C. 5^3

D. 6^2

8 A building contractor is using a wire and a pulley to lift materials to the roof of a building.

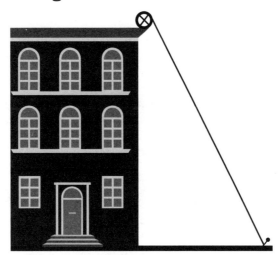

The contractor used the Pythagorean theorem to determine that the length of the wire is $7\sqrt{3}$. Which of the following numbers is closest to the length of the wire?

A. 12 feet

B. 18 feet

C. 21 feet

D. 28 feet

9 Which value is closest to 0.80?

A. $\dfrac{1}{3}$

B. $\dfrac{2}{3}$

C. $\dfrac{4}{5}$

D. $\dfrac{6}{7}$

10 **Which equation illustrates the associative property of multiplication?**

A. $f + g + h = fgh$

B. $(fg)h = f(gh)$

C. $(fg) + h = f + (gh)$

D. $\dfrac{fg}{h} = \dfrac{f}{gh}$

11 **The equation $(2)(3x + 4) = (3x + 4)(2)$ is true for all real numbers. Which property does this exemplify?**

A. associative property of addition

B. association property of multiplication

C. commutative property of multiplication

D. commutative property of addition

Chapter 1 Answer Explanations

Let's Review 1: Equivalent Numbers

1. C

The square root of 21 is approximately 4.583. Therefore, the best answer choice is 4.6.

2. A

If you press the square root key on your calculator and then 144, you'll see that the square root of 144 is 12. So, 12^2 is the correct answer. Or you can see that A is the correct answer by pressing the x^2 key and then 12, which will give you 144.

3. B

You can answer this question two different ways. You can use your calculator to divide the denominator of each fraction into the numerator. You can also answer this problem by converting .45 into a fraction. To do this, move the decimal point two places to the right and put 45 over 100. Then reduce the fraction (the common factor is 5).

4. 25%

To convert a decimal to a percent, move the decimal point two places to the right.

5. C

The square root of 32 is about 5.6. Point C is closest to this number.

Let's Review 2: Equivalent Expressions

1. C

According to the associative property, numbers can be grouped in any order. Answer choice C shows the numbers grouped in a different order. Choices A and D involve multiplication. Choice B is not right because 92 is subtracted instead of added.

2. C

The additive inverse is an opposite number, which when added to the original number equals zero. The additive inverse of -24 is 24 because when they are added, the sum is zero.

3. A

The commutative property states that numbers can be multiplied in any order. Answer choice A is the correct answer.

4. C

According to the inverse property of multiplication, a number should be multiplied by another number (a fraction) that will equal 1.

5. D

The distributive property applies to both addition and multiplication.

Chapter 1 Practice Problems

1. C

To solve this problem, you can eliminate answer choices A and B, since they are obviously smaller than the numbers in answer choices C and D (the numbers as well as the exponents are smaller). Use your calculator to find the correct answer between choices C and D.

2. C

Remember that to convert a decimal into a percent, move the decimal point two places to the right and add the percentage sign.

3. A

The number 4^{-3} multiplied by 4^3 is 1, since this is its opposite. Remember that 4^{-3} means $\frac{1}{4^3}$.

4. A

According to the associative property of addition, you can add numbers in any order and the sum is the same.

5. C

You might be able to answer to this problem by sight. Answer choice C, 1.21×10^3, is the greatest.

6. C

According to the identity property of multiplication, any number multiplied by 1 is the same number.

7. C

The number 5^3 is 125, and this is the greatest number.

8. A

The square root of 3 is about 1.7. If you multiply this by 7, it is about 12.

9. C

If you convert the decimal 0.80 to a fraction, you get $\frac{80}{100}$. When you reduce this number, you get $\frac{4}{5}$.

10. B

The associative property of multiplication states that numbers can be multiplied in any order. Answer choice B is an example of this property.

11. C

The commutative property states that numbers can be added or multiplied in any order. In the equation $(2)(3x + 4) = (3x + 4)(2)$, think of $3x + 4$ as representing one number, say 13. $(2)(13) = (13)(2)$ shows the commutative property. Answer choice C is correct.

Chapter 2

Number Sense and Operations, Part 2

$$6^2$$
$$35\%$$
$$87(3)$$
$$19x-2$$

Standards

10.N.2 Simplify numerical expressions, including those involving positive integer exponents or the absolute value, for example, $3(24 - 1) = 69$, $4\,|3 - 5| + 6 = 14$; apply such simplifications in the solution of problems.

10.N.4 Use estimation to judge the reasonableness of results of computations and of solutions to problems involving real numbers.

Some Number Sense and Operations questions on the MCAS will ask you to estimate an answer. You may be asked to choose the best estimate of the sum or difference of a set of numbers. You may be asked to look at an illustration and estimate an amount shown. Most questions will be about real-life situations such as having to estimate a discount for an item on sale.

You will also be asked to solve simple expressions that do not contain a variable, such as x or y. Often you will have to raise some numbers to a power or find the absolute value of an expression. Review these concepts in Chapter 1 if you're not sure how to do such problems.

Other questions on the MCAS will be about money. For these questions, you may be asked to figure out how much something costs after a discount, or you may be asked to determine a discount, such as 25% off a $100 item. Other questions involving money might ask you to calculate interest and sales tax. You may use a calculator only to answer questions in Session 2.

Estimation

When you **estimate**, you find the approximate value. You'll learn some of the most common methods of estimation in this chapter.

Rounding

To approximate answers on the MCAS, you'll have to round numbers to the same place value. The following numbers are rounded to the tens place:

$$12 + 19$$
$$10 + 20$$

If you add the estimations of these numbers, the answer is 30. The following numbers are rounded to the hundreds place:

$$186 + 342$$
$$200 + 300$$

The estimated answer is 500. If you rounded the numbers to the tens place, they would look like this:

$$190 + 340$$

Can you estimate this number in your head? If not, use a pencil and paper. The estimated answer is 530.

These numbers are rounded to the thousands place:

$$1,230 + 4,689$$
$$1,000 + 5,000$$

The estimated answer is 6,000. If you rounded these numbers to the hundreds place, they would look like this:

$$1,200 + 4,700$$

The estimated answer is 5,900.

Round each of the numbers to the tens place to solve this problem:

Estimate the sum of 42, 14, 28, 23, and 29.

If you round the numbers to the tens place, they look like this:

Number	Rounded to the Tens Place
42	40
14	10
28	30
23	20
29	30

Now add the rounded numbers:

$$40 + 10 + 30 + 20 + 30 = 130$$

The estimated answer is 130.

Front-End Estimation

Another type of rounding is **front-end estimation**. With this type of estimation, you round and add only the numbers in the leftmost place. Front-end estimation was used to estimate the difference or sum of the numbers below:

$$45,736 - 28,924$$
$$50,000 - 30,000 = 20,000$$

$$154 + 790$$
$$200 + 800 = 1,000$$

$$1,241 + 3,880$$
$$1,000 + 4,000 = 5,000$$

Use front-end estimation to solve this problem:

Estimate the difference of 2,945 − 1,523.

If you use front-end estimation, the numbers are rounded as follows:

Number	Rounded Using Front-End Estimation
2,945	3,000
1,523	2,000

Now use the rounded number to estimate the difference:

$$3,000 - 2,000 = 1,000$$

See how it works? Other estimation questions on the MCAS will ask you to estimate parts of a whole. For example, look at the figure below.

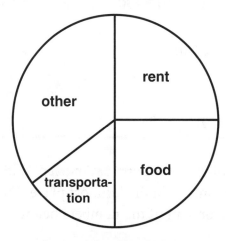

Mason Family Budget

About how much of its budget does the Mason family spend on rent?

By looking at the circle, you can estimate that the family spends about one quarter, or 25%, of its budget on rent.

Let's Review 3: Estimating Numbers

Complete each of the following questions. Use the Tip below each question to help you choose the correct answer. When you finish, check your answers with those at the end of Chapter 2.

1 Estimate the sum of 32, 15, 67, 99, and 13.

A. 200

B. 210

C. 230

D. 240

 TIP

> Round 32, 15, 67, and 13 to the nearest ten.
> Round 99 to 100.

2 Sarah earns $7 an hour bagging groceries at a corner store during the 10 weeks of summer vacation. If she averages 20 hours per week, what is a reasonable estimate of what Sarah will earn during the summer?

A. $140

B. $170

C. $1,400

D. $14,000

 TIP

> Begin by multiplying 7 by 20.

3 Look at the figure below, which shows the seating in a theater.

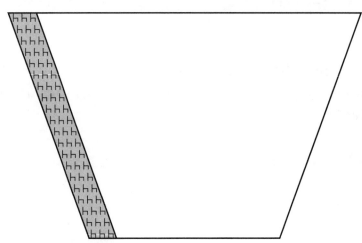

If the shaded area represents 100 seats, about how many seats are in the theater?

A. 100

B. 1,000

C. 10,000

D. 100,000

About how many shaded areas would fit into the diagram of the theater?

4 Estimate the difference: 32,987 − 12,956

A. 10,000

B. 20,000

C. 30,000

D. 40,000

Round both numbers to the nearest ten thousand. Then subtract.

Numerical Expressions

Some questions on the MCAS will ask you to simplify expressions. As you'll learn later in this book in the chapters about algebra, an **expression** is a relationship between numbers. You may be able to simplify an expression, but you can't solve it. An **equation**, on the other hand, contains an equal sign and can be solved. Note that the expressions tested under the Number Sense and Operations standards do not involve variables, such as *x* and *y*. Expressions with variables are tested under Patterns, Relations, and Algebra standards and will be discussed later in this book.

Expression: 6(2 + 4)
Simplified: 36

Let's Review 4: Numerical Expressions

Complete each of the following questions. Use the Tip below each question to help you choose the correct answer. When you finish, check your answers with those at the end of Chapter 2.

1   What is the value of the following expression?

$$\frac{14(2 + 3)}{2(3 + 2)}$$

A.   7

B. $\frac{35}{144}$

C. $\frac{35}{10}$

D.   10

TIP

Perform the operations in parentheses first.

2   **Which of the following is equivalent to (25 − 10)²?**

A. $|10 - 25|^2$

B. $-(25 - 10)^2$

C. $10^2 - 25^2$

D. $-|25 - 10|^2$

> **TIP**
>
> Remember that the absolute value of a negative number is its opposite.

3   **What is the value of the following expression?**

$$|-8 + 2| - |2^2 - 2|$$

A. -4

B. -2

C. 4

D. 8

> **TIP**
>
> The absolute value of a negative number is its opposite, meaning the absolute value of −6 is 6.

Computing Money

Some questions on the MCAS will be about adding, subtracting, dividing, and multiplying monetary amounts. These questions are about real-life situations. You might be asked to calculate the sale price of a discounted item, sales tax, or the interest on a short-term loan.

Discounts and Sale Prices

Some questions on the MCAS will ask you to determine the amount of a discount for an item on sale or the sale price of an item. For example, you might be asked to determine a discount of 15% on shoes that cost $45.00. To do this with paper and pencil, you would multiply 45 by .15 as shown here:

$$
\begin{array}{r}
45 \\
\times.15 \\
\hline
225 \\
+450 \\
\hline
6.75
\end{array}
$$

The discount is $6.75.

Other questions will ask you to determine the sale price of an item after a discount is applied. Read the following problem.

Megan wants to buy a mirror for her room that is usually priced at $85.00 and is now discounted by 40%. What is the sale price of the mirror?

To solve this problem using a pencil and paper, you would multiply .40 by 85 as shown here:

$$
\begin{array}{r}
85 \\
\times.40 \\
\hline
00 \\
34.00
\end{array}
$$

Remember that $34 is the amount of the discount. This question asks you to find the sale price of the mirror, so you have to subtract 34 from 85:

$$
\begin{array}{r}
85.00 \\
-34.00 \\
\hline
51.00
\end{array}
$$

The sale price of the mirror is $51.00.

You determine the sales tax on an item in much the same way that you determine a discount. Read this question:

Gail works in a small hardware store where the cash register does not compute the sales tax. If the sales tax is 7%, what amount should Gail add to a purchase of $10.00?

To answer this question, multiply 10 by .07 or 7%. The amount of sales tax Gail should add to a purchase of $10.00 is $0.70 or 70 cents.

Some questions might ask you to add the sales tax to the cost of an item. Read this problem:

Brian wants to buy a bike that costs $125. He knows that he will have to pay 6% sales tax on the bike. How much money, including tax, does Brian need to buy the bike?

To answer this question, you have to calculate the sales tax and add it onto the cost of the bike. Multiply .06 or 6% by 125. When you do this, you get $7.50. Now add this amount to $125, the cost of the bike. The answer is $132.50. This is the amount of money Brian needs to buy the bike.

Interest

When you borrow money, you take a loan. Usually, you're asked to pay interest on the loan. **Interest** is an additional sum of money you must pay in addition to the **principal**, the amount of money you borrowed. Interest is like a fee that you pay to the person or company that lent you the money. On the MCAS, calculating interest is usually very simple. Read this problem:

If Alberto borrows $5,000 from a bank at a fixed interest rate of 12% per year, how much interest must he pay if he pays the loan in full at the end of one year?

To solve this problem, you must multiply .12 or 12% by 5,000. The answer is $600. If Alberto pays the loan in full at the end of one year, he must pay $600 in interest.

Matrices

You may be asked questions on the MCAS involving matrices. These questions may or may not have variables, which are letters representing unknown values. Use the following rules when adding, subtracting, or dividing matrices.

Addition and Subtraction

To add or subtract matrices, add or subtract the corresponding members, paying close attention to the negative signs. Remember that a double negative is positive.

$$\begin{bmatrix} -4 & 0 \\ 3 & 1 \end{bmatrix} + \begin{bmatrix} 4 & -3 \\ 3 & 2 \end{bmatrix}$$

$$\begin{bmatrix} (-4+4) & (0-3) \\ (3+3) & (1+2) \end{bmatrix}$$

$$\begin{bmatrix} 0 & -3 \\ 6 & 3 \end{bmatrix}$$

Multiplication

A matrix can be multiplied by a number or another matrix. In this example, the matrix is multiplied by a number:

$$2\begin{bmatrix} -8 & 0 \\ 2 & 3 \end{bmatrix}$$

$$=\begin{bmatrix} -16 & 0 \\ 4 & 6 \end{bmatrix}$$

In this example, two matrices are multiplied:

$$\begin{bmatrix} 2 & 1 \\ 3 & -1 \end{bmatrix} \times \begin{bmatrix} 3 \\ 4 \end{bmatrix} = \begin{bmatrix} (2)(3)+(1)(4) \\ (3)(3)+(-1)(4) \end{bmatrix} = \begin{bmatrix} 10 \\ 5 \end{bmatrix}$$

Let's Review 5: Computing Money

Complete each of the following questions. Use the Tip below each question to help you choose the correct answer. When you finish, check your answers with those at the end of Chapter 2.

1 Mario wants to buy a skateboard that is regularly priced at $55 but is now discounted by 15%. What is the sale price of the skateboard?

A. $8.25

B. $46.75

C. $56.75

D. $82.50

Multiply 55 by .15. Then deduct this amount from the price of the skateboard.

Short-Response Item

2 Javier works in an ice cream store where the cash register does not compute the sales tax. If the sales tax is 5%, what is the amount Javier should add to a purchase of $11.00?

Multiply .05 or 5% by 11.

3 If a pair of jeans originally cost $25 and is selling at a 12% discount, what is the amount of this discount?

A. $3.20

B. $3.00

C. $23.00

D. $30.80

TIP

Multiply 25 by .12. This is the amount of the discount.

Short-Response Question

4 Leo's dental plan pays 45% of dental expenses after the deductible of $100 is subtracted. Leo's total dental bill was $380. What is the exact amount the insurance company will pay? Show all work.

TIP

The deductible is the customer's initial out-of-pocket expense before the insurer begins payments.

5 If Javier borrows $7,000 to buy a car at a fixed simple interest rate of 13% per year, how much interest must he pay if he pays the loan in full at the end of two years?

A. $910

B. $920

C. $1,820

D. $1,840

TIP

Notice that this question asks you to determine the interest for TWO years. First find the interest for one year, and then multiply this number by two.

Short-Response Question

6 **Add these matrices:**

$$[-3 \quad 0] + [4 \quad -2]$$

Remember to add the corresponding members. Begin with −3 and 4.

Chapter 2 Practice Problems

Complete each of the following practice problems. Check your answers at the end of this chapter. Be sure to read the answer explanations!

1 Kate works in a small gift shop where the cash register does not compute the sales tax. If the sales tax is 6%, how much should Kate charge the customer in total for a purchase of $25.00?

A. $1.50

B. $1.75

C. $26.50

D. $41.50

Short-Response Question

2 What is the value of the following expression? Show all work.

$$3(5 + 7) + (2 + 2)^3$$

3 Kristen wants to buy a coat that is usually priced at $125 and is now discounted by 35%. What is the sale price of the coat?

A. $43.75

B. $81.25

C. $90.00

D. $91.50

Short-Response Question

4 Miguel earns $15 per yard to mow grass and he mows 3 lawns per week. How much does Miguel earn in 4 weeks?

5 Keisha's eyeglass plan pays 52% of the cost for a pair of glasses after the deductible of $25 is subtracted. Keisha's glasses cost $225. Which is the best estimate of the amount the insurance company will pay?

A. $90

B. $100

C. $120

D. $140

6 What is the value of the following expression?

$$|2 + -4| + |-(3^2) + 4|$$

A. -7

B. 0

C. 5

D. 7

7 If boots that originally cost $52 are selling at a 25% discount, what is the amount of the discount?

A. $12

B. $13

C. $27

D. $39

8   If Karen borrows $8,000 from a bank at a fixed interest rate of 14% per year, how much interest must she pay if she pays the loan in full at the end of one year?

A. $1,120

B. $2,120

C. $5,880

D. $6,880

9   Pedro wants to buy a birthday present for his mother that costs $32.00. He knows he must pay 7% sales tax on the gift. What amount should Pedro add to a purchase of $32.00?

A. $2.00

B. $2.24

C. $22.40

D. $29.76

10 **Look at the figure below.**

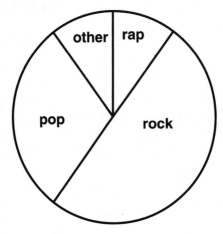

About how many students chose rock as their favorite kind of music?

A. 10%

B. 20%

C. 30%

D. 50%

Chapter 2 Answer Explanations

Let's Review 3: Estimating Numbers

1. C

The numbers should be rounded to the nearest ten: 30, 20, 70, and 10, and 99 should be rounded to 100. When you add these numbers, you get 230.

2. C

Sarah works an average of 20 hours a week, so you need to multiply $7 by 20. The answer is $140, the amount Sarah earns each week. Then you need to multiply $140 by 10, the number of weeks Sarah works. The answer is $1,400.

3. B

By looking at the picture, you can estimate that there are about 10 sections in the illustration. If each section is about 100 seats, there are about 1,000 seats.

4. B

If you round both numbers to the nearest ten thousand, you get 30,000 and 10,000. The difference between these numbers is 20,000.

Let's Review 4: Numerical Expressions

1. A

To simplify this expression, perform the operations in parentheses: 14 (5)/2 (5). Simplify further: $\frac{70}{10}$ or 7.

2. A

$(25 - 10)^2 = 15^2 = 225$. Answer choice A is correct because the absolute value of $10 - 25$ is $+15$ and $(+15)^2 = 225$. Answer choice B is -225, answer choice C is -525, and answer choice D is -225.

3. C

To solve this expression, begin by solving the expression inside the absolute value signs and then take the opposite: $|-8+2| = |-6|$ and $|2^2 - 2| = |4 - 2| = |2|$. Then remove the absolute value signs and simplify the expression: $6 - 2 = 4$.

Let's Review 5: Computing Money

1. B

When you multiply $55 by .15, you get $8.25. When you subtract this amount from the original cost of the skateboard, you get $46.75.

2. $0.55

To find the amount of the sales tax, you need to multiply $11 by .05. The answer is $0.55.

3. B

To solve this problem, you have to multiply $25 by .12. The amount of the discount is $3.00.

4. $126

The first step in solving this problem is to subtract 100 from 380. Then multiply .45 by 280. The answer is $126. This is the amount Leo's insurance will pay.

5. C

To solve this problem, you have to multiply $7,000 by .13. The answer is $910. However, since you need to find the amount of interest Javier would owe in two years, you then need to multiply $910 by 2.

6.

$$[(-3 + 4) \quad (0 + -2)]$$
$$= [1 - 2]$$

Chapter 2 Practice Problems

1. C

To solve this problem, you need to multiply $25 by .06. When you do this, you get $1.50. Then add this amount to $25. The answer is $26.50.

2. 100

Begin by performing the operations in parentheses: $3(12) + (4)^3$. Then multiply and add the resulting products: $36 + 64 = 100$.

3. B

To solve this problem, multiply 125 by .35. Then deduct this amount from the original price of the coat.

4. $180

If you multiply $15 by 3, you get $45. If you multiply $45 by 4, you get $180.

5. B

If you deduct $25 from $225, you get $200. If you multiply $200 by .52, you get $104. The best estimate is B, $100.

6. D

First, perform the operations inside the absolute value signs: $|-2| + |-5|$. Then find the absolute value: $2 + 5 = 7$.

7. B

If you multiply the price of the boots, $52, by the amount of the discount, .25, you get $13.

8. A

To solve this problem, you have to multiply $8,000, the amount of money Karen borrowed, by .14, the amount of the interest. The answer is $1,120.

9. B

To find the sales tax on $32, multiply this number by .07, the amount of the sales tax. The answer is $2.24. This is the amount that Pedro should add to the cost of the purchase.

10. D

About half of all students chose rock as their favorite music.

Chapter 3

Data Analysis, Statistics, and Probability, Part 1

Standards

10.D.1   Use appropriate statistics (e.g., mean, median, range, and mode) to communicate information about the data. Use these notions to compare different sets of data.

10.D.3   Describe and explain how the relative sizes of a sample and the population affect the validity of predictions from a set of data.

In this chapter, you'll learn how to solve data analysis problems involving probability and measures of central tendency. **Probability** is the odds (or probability) that an event will happen. Like many questions on the MCAS, probability problems will often involve real-life situations.

The **statistical quantities** tested on the MCAS include mean, mode, median, and range. You'll learn how to determine each of these for a set of a data.

Probability

Probability can be determined using this formula:

P = number of favorable outcomes/number of possible outcomes

Probability can be expressed as a fraction, a decimal, or a ratio. Most of the time on the test, probability is expressed as a fraction.

Read this problem:

Find the probability of spinning a 3 on the spinner below.

A. 0

B. $\dfrac{1}{4}$

C. $\dfrac{1}{2}$

D. 1

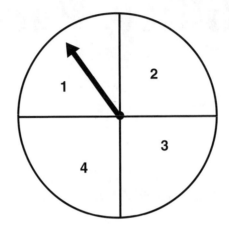

To solve this problem, use the formula shown above.

$1 =$ the number of favorable outcomes

$4 =$ the number of possible outcomes

Using this formula, you can see that the probability of spinning a 3 on the spinner is $\dfrac{1}{4}$. Answer choice B is correct. Let's try another problem:

Justine has a bag of twenty marbles. Ten of these marbles are white, 3 are green, 2 are blue, and 5 are yellow. If Justine reaches into the bag and pulls out a marble without looking, what is the probability that she will pull out a yellow marble?

A. 0

B. $\dfrac{1}{20}$

C. $\dfrac{1}{4}$

D. $\dfrac{1}{2}$

Use the probability formula to solve this problem. There are 5 yellow marbles, so this is the number of favorable outcomes. There are 20 marbles altogether, so this is the number of possible outcomes. The probability that Justine will pull out a yellow marble is $\frac{5}{20}$. Answer choice C is correct, since $\frac{5}{20}$ reduces to $\frac{1}{4}$.

You might have to solve a probability problem on the MCAS involving a tree diagram. Read this problem:

Use the tree diagram to predict the probability of flipping one coin four times and getting all heads or all tails.

A. $\frac{1}{16}$

B. $\frac{1}{8}$

C. $\frac{1}{4}$

D. $\frac{1}{2}$

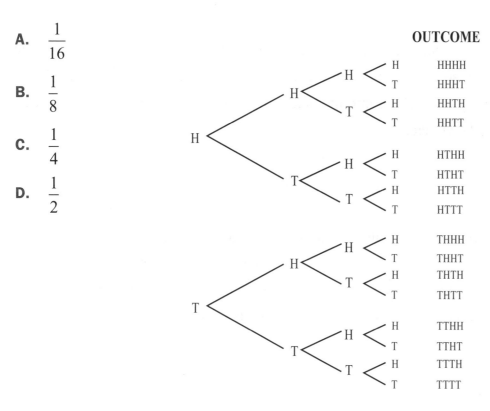

Look carefully at the tree diagram. The letter H represents heads, and the letter T represents tails. If you look at these letters, you'll see that there are sixteen possible outcomes. Two of these outcomes are HHHH and TTTT, meaning that the coin came up all heads or all tails. If you use the probability formula, you can see that the probability is $\frac{2}{16}$ or $\frac{1}{8}$. Answer choice B is correct. Let's try one more problem:

There are 10 straws in a box; some are white and some are red. The probability of reaching into the box and selecting a white straw is $\frac{2}{5}$. What is the probability of selecting a red straw?

A. $\dfrac{1}{10}$

B. $\dfrac{3}{5}$

C. $\dfrac{4}{5}$

D. 1

You have to work backward to solve this problem. You know there are 10 straws in the box altogether, so the denominator must be 10 before it is reduced. You also know that the probability of reaching into the box and pulling out a white straw is $\frac{2}{5}$. Set up the fractions as shown here:

$$\frac{2}{5} = \frac{x}{10}$$

Ten divided by 5 is 2, so you need to multiply the numerator in the first fraction, 2, by 2. The answer is $\frac{4}{10}$. If 4 out of the 10 straws are white, then 6 out of the 10 straws are red. The fraction $\frac{6}{10}$ reduced is $\frac{3}{5}$. The probability of reaching into the box and pulling out a red straw is $\frac{3}{5}$, so answer choice B is correct.

Let's Review 6: Probability

Complete each of the following questions. Use the Tip below each question to help you choose the correct answer. When you finish, check your answers with those at the end of Chapter 3.

1   Use the diagram to predict the probability that a family with three children has three girls or three boys.

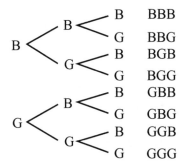

A. $\dfrac{1}{4}$

B. $\dfrac{1}{3}$

C. $\dfrac{1}{2}$

D. 1

Count the number of combinations, using the letters to the right of the tree diagram. Notice that out of all of the combinations, two are GGG and BBB, meaning all girls and all boys.

2 Find the probability of spinning "green" on the spinner below.

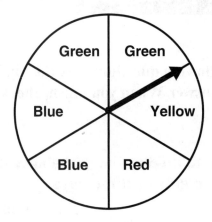

A. 0

B. $\frac{1}{4}$

C. $\frac{1}{3}$

D. $\frac{1}{2}$

Remember to use the formula for probability and then reduce the fraction. There are six sections on the spinner, and two of these sections are green.

3 A bag contains 8 white chips, 5 red chips, 3 black chips, 2 blue chips, and 2 green chips. If you reach into the bag without looking, what is the probability that you will pull out a red chip?

A. $\frac{1}{4}$

B. $\frac{1}{3}$

C. $\frac{2}{3}$

D. 5

There are 20 chips altogether and 5 of these chips are red.

4   Peter is going to roll a six-sided number cube. What is the probability of rolling an even number?

A. $\dfrac{1}{6}$

B. $\dfrac{1}{4}$

C. $\dfrac{1}{3}$

D. $\dfrac{1}{2}$

TIP

A six-sided number cube has sides numbered 1, 2, 3, 4, 5, and 6.

 5   There are eight jelly beans in a jar; some are pink and some are yellow. The probability of randomly reaching into the jar and selecting a pink jelly bean is $\dfrac{1}{4}$. What is the probability of reaching into the jar and pulling out a yellow jelly bean?

A. $\dfrac{1}{4}$

B. $\dfrac{1}{2}$

C. $\dfrac{3}{4}$

D. $\dfrac{3}{5}$

TIP

Remember that the fraction is reduced. To find out how many yellow jelly beans are in the jar, you need a denominator of 8.

 6   If a penny is tossed 10 times and on the first five tosses it comes up heads, what is the probability of getting heads on the sixth toss?

A. $\dfrac{1}{4}$

B. $\dfrac{1}{3}$

C. $\dfrac{1}{2}$

D.   1

TIP

When you toss a coin, the odds that it will come up either heads or tails are the same regardless of how many times you have tossed it.

Mean, Median, Mode, and Range

Some questions on the MCAS will ask you to analyze data to find statistical quantities, including the mean, median, mode, and range. **Mean** is another word for "average." To find the mean of a set of numbers, add all of the numbers together and divide by the number of items that make up that total. Look at this set of numbers:

2, 4, 6, 8, 10

To find the mean, you first add all of the numbers:

$$2 + 4 + 6 + 8 + 10 = 30$$

Then you divide 30 by the number of items, in this case 5. The mean of these numbers is 6.

The **median** of a set of numbers is the middle number when the numbers are ordered by size. It's not the average, but simply the number in the middle. Look at this set of numbers:

10, 4, 2, 8, 6

To find the median, you need to put them in order from least to greatest:

2, 4, 6, 8, 10

When the numbers are in order from least to greatest, you can see that the number 6 is the median.

The **mode** of a set of data is the most frequently occurring number. Look at the numbers below:

88, 90, 76, 42, 88, 92, 100, 110, 115

The mode of these numbers is 88, the only number that occurs more than once.

The **range** of a set of data is the difference between the smallest number and the largest number. Consider these numbers again:

$$88, 90, 76, 42, 88, 90, 100, 110, 115$$

The smallest number is 42 and the largest is 115. To find the range, subtract 42 from 115:

$$115 - 42 = 73$$

The range of this set of numbers is 73.

Now read this short-response problem, the kind of problem about central tendency that you might see on the MCAS:

The scores on Mr. Seymour's English test were 98, 60, 88, 87, 96, 79, 80, 58, 76, 99, 80, 58, 76, 99, 90, 87, 62, 76, 89, and 97. What is the range of the scores?

To determine the range, subtract the lowest test score, 58, from the highest, 99.

$$99 - 58 = 41$$

Forty-one is the correct answer. Now let's try this problem:

If the mean number of people who attended an amusement park over 5 days is 25,000, what was the total attendance during the 5 days?

A. 5,000

B. 50,000

C. 125,000

D. 250,000

To solve this problem, you need to multiply the number of days by the mean. In this case, you would multiply 5 × 25,000. The answer is C, 125,000.

Let's Review 7: Mean, Median, Mode, and Range

Complete each of the following questions. Use the Tip below each question to help you choose the correct answer. When you finish, check your answers with those at the end of Chapter 3.

1   The office manager in a small office is considering hiring a receptionist to answer the telephone. To see whether or not a receptionist is needed, the employees used a log to record the number of calls answered each day. The data recorded in a 14-day period is shown below.

10, 12, 8, 16, 6, 5, 8, 5, 12, 13, 12, 12, 8, 6

Which of the following has the highest value?

A.   mean

B.   mode

C.   median

D.   range

TIP

The median of an even sample of data equals the mean of the middle two numbers.

2   The number of cars sold at Ray's Used Automobiles was 12 in January, 22 in February, 30 in March, 42 in April, and 58 in May. What is the range in the number of cars sold from January to May?

A.   12

B.   33

C.   46

D.   70

TIP

Remember that the range is the difference between the least and the greatest.

3 The total points scored for the Warriors basketball team for each game during the season were 42, 20, 13, 64, 27, 35, 45, 40, 23, 12, 12, and 39. What is their mean score?

A. 12

B. 31

C. 35

D. 52

TIP

You need to find the mean of this data. Add the numbers, and then divide by the total number of items.

4 Tickets sold at Central Elementary are shown below.

Grade	Number of Tickets Sold
1	246
2	112
3	493
4	98
5	209
6	112
7	190

Use the data in the chart to determine the median.

A. 98

B. 112

C. 190

D. 209

TIP

Put the number of tickets sold in order from least to greatest. The median is the number in the middle.

Open-Response Question

5 Renee's World Cultures grades were 84, 85, 95, 88, 92, 100, 82, and 84.

a. What is the mean of Renee's grades? Show your work.

b. What is the range of her grades?

c. Renee would like her average to be a 90. What is the minimum grade she must score on the next test to achieve this goal?

Chapter 3 Practice Problems

Complete each of the following practice problems. Check your answers at the end of this chapter. Be sure to read the answer explanations!

1 There are 12 coins in a box; some are nickels and some are pennies. The probability of randomly reaching into the box and pulling out a nickel is $\frac{2}{3}$. What is the probability of reaching into the box and pulling out a penny?

A. $\frac{1}{8}$

B. $\frac{1}{4}$

C. $\frac{1}{3}$

D. $\frac{2}{3}$

Short-Response Question

2 The weekly salaries of seven employees are $160, $240, $260, $85, $200, $180, and $120. What is the median salary?

3 Find the probability of spinning a 4 on the spinner below.

A. $\frac{1}{8}$

B. $\frac{1}{4}$

C. $\frac{1}{3}$

D. $\frac{1}{2}$

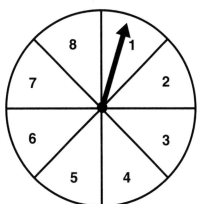

4 Use the tree diagram to predict the probability of flipping a coin and getting heads and rolling a number cube and getting an odd number.

A. $\dfrac{1}{12}$

B. $\dfrac{1}{4}$

C. $\dfrac{1}{3}$

D. $\dfrac{1}{2}$

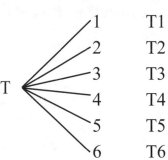

```
        1   H1
        2   H2
        3   H3
   H    4   H4
        5   H5
        6   H6

        1   T1
        2   T2
        3   T3
   T    4   T4
        5   T5
        6   T6
```

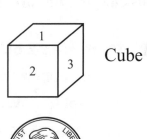

Cube

Coin

5 The average annual temperature in Sarasota, Florida, is shown in the table below.

Month	Average Temperature
January	72°
February	74°
March	78°
April	82°
May	87°
June	90°
July	91°
August	91°
September	90°
October	85°
November	79°
December	74°

Use the chart to determine the range.

A. 19°

B. 20°

C. 83°

D. 91°

6 Christine has a bag of 25 marbles. Five of these marbles are green, 4 are blue, 3 are white, 8 are black, and 5 are yellow. If Christine reaches into the bag, what is the probability that she will randomly pull out a yellow marble?

A. $\frac{1}{8}$

B. $\frac{1}{5}$

C. $\frac{1}{4}$

D. $\frac{1}{3}$

7 Kayla recorded the hours she spent studying each week for five weeks. She listed these hours in the chart below.

Week	Hours Spent Studying
1	18
2	12
3	15
4	5
5	10

What is the mean number of hours Kayla spent studying?

A. 10

B. 12

C. 13

D. 15

8 The miles Michelle ran in a week are 2, 3, 3, 4, 4, 3, and 6. What is the mode number of miles Michelle ran?

A. 3

B. 4

C. 5

D. 6

9 The youngest person in an audience of 300 people is 15 years old. The range of ages is 52 years. What is the age of the oldest member of the audience?

A. 52

B. 58

C. 67

D. 72

Chapter 3 Answer Explanations

Let's Review 6: Probability

1. A

If you look at the letters to the right of the diagram, you can see that there are eight possibilities, or combinations of children. Out of these eight, there is one possibility that the three children will be all girls, and one possibility that the three children will be all boys. So the probability that the children will be either all girls or all boys is $\frac{2}{8}$. When you reduce this fraction, you get $\frac{1}{4}$.

2. C

The spinner is divided into six sections and two of these sections are green. So the probability that the spinner will land on green is $\frac{2}{6}$ or $\frac{1}{3}$.

3. A

The bag contains 20 chips altogether and 5 of these chips are red. The probability of pulling out a red chip is $\frac{5}{20}$ or $\frac{1}{4}$.

4. D

If the number cube has six sides and is numbered 1, 2, 3, 4, 5, and 6, three of the six sides are even. Therefore, the probability of rolling an even number is $\frac{1}{2}$.

5. C

The question tells you that there are 8 jelly beans in a jar, so the denominator should be 8. The question also says that the probability of selecting a pink jelly bean is $\frac{1}{4}$ or $\frac{2}{8}$. If you subtract 2 from 8, the total number of jelly beans, you get 6. So the probability of choosing a yellow jelly bean is $\frac{6}{8}$ or $\frac{3}{4}$.

6. C

If you toss a penny, the odds that it will come up heads or tails are always $\frac{1}{2}$, regardless of how many times you toss the penny.

Let's Review 7: Mean, Median, Mode, and Range

1. B

To solve this problem, you need to calculate each measure of central tendency and choose the measure showing the greatest number of calls. The mean is 9.5, the mode is 12, the median is 9 (the average of 8 and 10), and the range is 11. Therefore, the mode shows the greatest number of calls.

2. C

To find the range in the number of cars sold from January to May, subtract the smallest number, 12, from the greatest number, 58. The range is 46.

3. B

When you add all of the numbers, you get 372. When you divide 372 by the number of items, 12, you get 31.

4. C

If you put the numbers in order from least to greatest, you'll see that 190 is in the middle. This number is the median.

5.

(a) 88.75

If you add together all of Renee's grades, you get 710. When you divide this number by 8, you get 88.75.

(b) 18

Renee's highest grade is 100. Her lowest grade is 82. To find the range, you must subtract 82 from 100. The answer is 18.

(c) 100

Renee needs a 100 in order to have an average of 90. Renee has a total of 710 for the 8 test grades. In order to have an average of 90 for 9 test grades, she must accumulate a total of $(9)(90) = 810$ points. $810 - 710 = 100$ points are needed on the next test.

Chapter 3 Practice Problems

1. C

The question tells you that there are 12 coins in a box and some are nickels and some are pennies, and the probability of randomly reaching into the box and pulling out a nickel is $\frac{2}{3}$. You know that the denominator must be 12, so the probability of pulling out a nickel before you reduce the fraction is $\frac{8}{12}$. Therefore, the probability of pulling out a penny is $\frac{4}{12}$ or $\frac{1}{3}$.

2. 180

If you put the salaries in order from least to greatest, you'll see that $180 is the median salary.

3. A

The spinner has eight sections and only one section is numbered 4, so the probability of spinning a 4 is $\frac{1}{8}$.

4. B

There are 12 possibilities and 3 of them involve rolling heads and an odd number on the number cube. So the probability that this will happen is $\frac{3}{12}$ or $\frac{1}{4}$.

5. A

To find the range in the average annual temperature, subtract the lowest temperature, 72°, from the highest temperature, 91°.

6. B

There are 25 marbles in the bag and 5 of them are yellow. Therefore, the probability of choosing a yellow marble is $\frac{5}{25}$ or $\frac{1}{5}$.

7. B

When you add all of the hours Kayla spent studying, you get 60. When you divide 60 by 5, you get 12.

8. A

The mode is the most frequently occurring number. In this case, it is the number 3.

9. C

To answer this question, simply add the range to the age of the youngest person.

Chapter 4

Data Analysis and Probability, Part 2

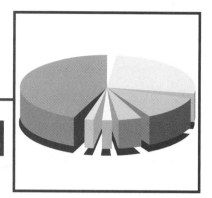

Standards

10.D.1 Select, create, and interpret an appropriate graphical representation (e.g., scatterplot, table, stem-and-leaf plots, box-and-whiskers plots, circle graph, line graph, and line plot) for a set of data.

10.D.2 Approximate a line of best fit (trend line) given a set of data (e.g., scatterplot). Use technology when appropriate.

In the previous chapter, you learned how to analyze a set of data to determine probability and measures of statistical quantity. In this chapter, you'll learn how to analyze data displayed in different forms. Some questions on the MCAS will be about data displayed in line, bar, and circle graphs. You need to be able to interpret data in these graphs to answer these questions. Other questions will be about data displayed in plots such as scatter plots, box-and-whisker plots, and stem-and-leaf plots. You'll learn about these graphs and plots in this chapter.

Line Graphs

A **line graph** is a very popular type of graph that compares two variables—one along the *x*-axis and one along the *y*-axis. Unlike in a bar graph, the two variables being compared in a line graph are closely related; a change in one variable causes a change in the second variable. A line graph is a great way to show trends. Look at this line graph:

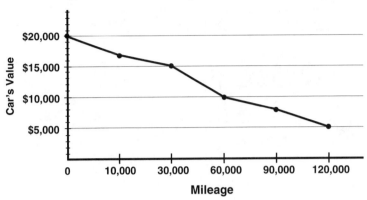

You can see from this line graph that as the number of miles on Tony's car increases, the value of the car decreases.

Bar Graphs

In a **bar graph**, the height or length of a bar shows the number of something. The higher or longer the bar, the greater the value. A bar graph has an *x*-axis and a *y*-axis, and is a good way to show comparisons. It can also show trends such as changes in sales over time.

While a bar graph can have either vertical or horizontal bars, most bar graphs on the MCAS have vertical bars. Look at this bar graph. It shows the number of cars manufactured at a factory over five years.

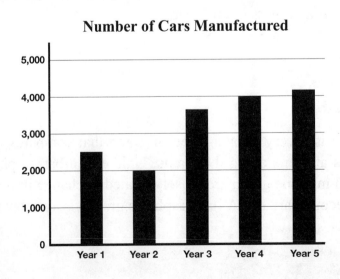

In this bar graph, Years 1 through 5 are shown on the *x*-axis and the number of cars manufactured is shown on the *y*-axis. You can see by just glancing at the graph that the greatest number of cars were manufactured in Year 5 and that except for Year 2, the number of cars manufactured increased each year.

Circle Graphs

A **circle graph**, also called a **pie chart** or **pie graph**, is often used to display the division of a whole or parts of a whole. Data in a circle graph is often displayed in percentages. Circle graphs work best to show large divisions such as the division of money in a household budget. Look at this circle graph:

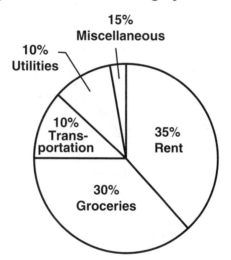

Miller Family Monthly Budget

You can see from this circle graph that the Miller family spends most of its monthly income on rent and groceries.

Venn Diagrams

Some questions on the MCAS may ask you to interpret data displayed in a Venn diagram. A **Venn diagram** is made up of two or more overlapping circles and is used to display relationships between two or more sets of data. Venn diagrams are a great way to show similarities and differences. Look at the Venn diagram below. If this diagram were filled in, it would compare two sets of data, A and B. The part of the circles that overlaps, C, would show ways that the sets of data are alike. Traits unique to each set of data would be in the part of the circles that do not overlap.

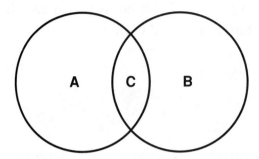

A Venn diagram comparing three sets of data would look like the one shown here. Note that the ways in which A and B are alike would be shown where circle A and circle B overlap. The ways in which A and C are alike would be shown where circle A and C overlap, and the ways in which B and C are alike would be shown where circle B and C overlap. The ways in which A, B, and C are alike would be shown in the small area where all three circles overlap.

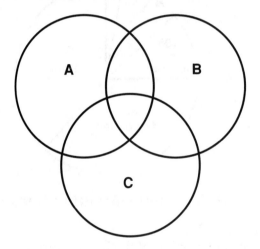

Let's Review 8: Graphs

Complete each of the following questions. Use the Tip below each question to help you choose the correct answer. When you finish, check your answers with those at the end of Chapter 4.

Extra-Credit Points in English

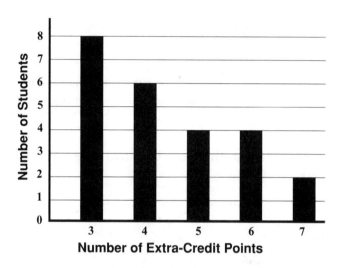

Number of Extra-Credit Points

1 The distribution of extra-credit points in Ms. Washington's English class is shown on the graph above. How many more students received five extra-credit points than seven extra-credit points?

A. 7

B. 3

C. 2

D. 5

TIP

To answer this question, look at the bar for five extra-credit points and the bar for seven extra-credit points. Subtract the number of students receiving seven extra-credit points from the number of students receiving five.

2 Alanis surveyed the students in her school to see what one thing they like to do in their spare time. About 45% of the students said they like to listen to music, 20% like to play sports, 22% like to read, and 13% like to do something else.

Which type of graph is appropriate to display the data Alanis collected in her survey?

A. line graph

B. bar graph

C. circle graph

D. Venn diagram

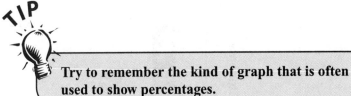

TIP

Try to remember the kind of graph that is often used to show percentages.

3 Bethany constructed a diagram to illustrate the number of students in her class that have a pet.

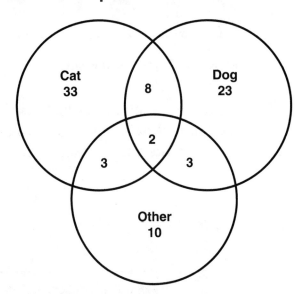

How many more students have a cat than a dog?

A. 8

B. 10

C. 20

D. 30

TIP

To answer this question, subtract the number of students with a dog from the number of students with a cat.

4 How many students have a dog, a cat, and another type of pet?

A. 2

B. 3

C. 8

D. 40

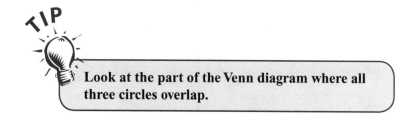

TIP

Look at the part of the Venn diagram where all three circles overlap.

Scatter Plots

A **scatter plot** shows at a glance whether there is a relationship between two sets of data. In a scatter plot, data is plotted by using dots. If the dots show a trend, a line is drawn. If the line is increasing from left to right, the trend is said to be positive. Look at the scatter plot shown here. It shows whether students who play a musical instrument get higher grades in school.

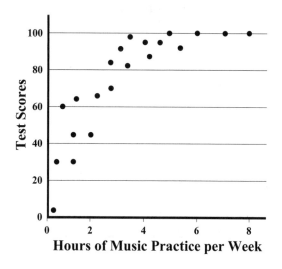

According to this scatter plot, students who play a musical instrument do earn higher grades. In fact, the more hours per week they practice playing an instrument, the higher their grades are.

Now look at the scatter plot below. It compares the same two variables, but this time it shows a negative trend.

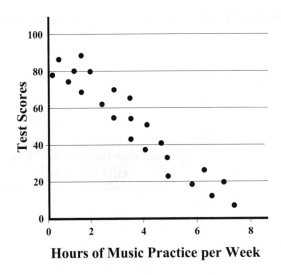

This scatter plot shows that students who play a musical instrument get lower grades. The more time they spend practicing, the lower their grades are.

Sometimes a scatter plot will show no trend between two variables. The scatter plot below shows no trend between playing a musical instrument and getting good grades.

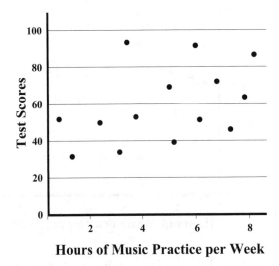

Box-and-Whisker Plots

On the MCAS, you might be asked to interpret data displayed in a box-and-whisker plot. A **box-and-whisker plot** is a great way to display large quantities of data. This type of plot looks like a box with two lines extending from it placed over a number line. In this type of plot, the lowest number in a set of data is shown by the leftmost horizontal line extending from the box, and the greatest number in a set of data is shown by the rightmost horizontal line. The median of the entire set of data is plotted with the vertical line in the center of the box. This is called the **second quartile**. Then the median of the data in between this median and the leftmost extreme is found. This is called the **first** (or lower) **quartile** and is shown by the left end of the box. In other words, the first quartile is the median of the lower part of the data. Then the median of the data in between this median and the rightmost extreme is found. This is called the **third** (or upper) **quartile** and is shown by the right end of the box. The third quartile is the median of the upper part of the data. Look at the box-and-whisker plot below and study each of its parts.

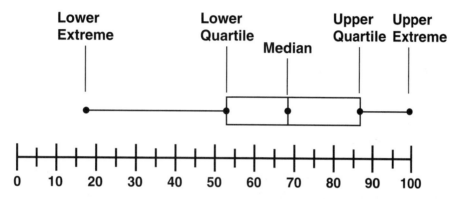

Stem-and-Leaf Plots

In a **stem-and-leaf plot**, each number in a set of data is split into a stem and a leaf. The first number goes into the stem column and the second goes into the leaf column. This type of plot makes it easy to categorize data and see a trend.

Imagine that a teacher wanted to create a stem-and-leaf plot to show these test scores: 99, 92, 86, 85, 74, 68, 71, 70, 85, 82, 83, 95, 80, 90. It would look like this:

Stem-and-Leaf	
6	8
7	0 1 4
8	0 2 3 5 5 6
9	0 2 5 9

By creating a stem-and-leaf plot, the teacher would be able to see the distribution of test scores.

Misleading Graphs

The MCAS may have a question that asks you to evaluate how a graph is biased or misleading. Something that is biased is unfair. Read this question:

> Javier wants to conduct a survey to see how many students
> are interested in joining a new school club.

Which sample population should Javier survey to represent the entire student body?

A. the teachers

B. the captain from each sports team

C. two people randomly from each homeroom class

D. 50 people randomly from the freshman class

To answer this question correctly, you need to choose the sample that is most fair. You can eliminate answer choice A, because surveying the teachers wouldn't give Javier an idea of how many students are interested in joining a new club. Answer choice B might be biased, since the captains of sports teams might enjoy the new club's activities if only they had the time. Answer choice C is a good answer, especially since Javier is randomly choosing two students. Answer choice D is biased, because students in the freshman class are all the same age and might not be a good representation of the whole school.

Graphs can also be misleading if the wrong kind of graph is used to display data. A bar graph, for example, should be used to show large differences in information. If it is used to show small differences, it will make them appear large, when in reality, they are not. The same is true of a line graph. It is not a good way to show small differences in information. Look at the graph on the next page.

Number of Minutes in which Students Ran Cross-Country Course

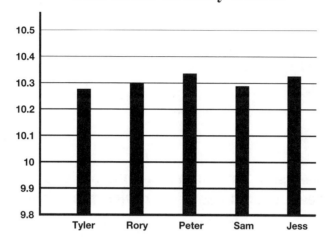

This bar graph is misleading. It makes very tiny amounts of time—seconds—look very large. Many of the bars also look the same. A simple table is a better way to display this information.

Let's Review 9: Plots

Complete each of the following questions. Use the Tip below each question to help you choose the correct answer. When you finish, check your answers with those at the end of Chapter 4.

Short-Response Question

1 The box-and-whisker plot shows the number of points Tyler's basketball team scored per game during the basketball season.

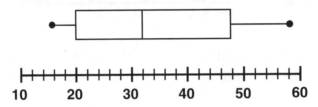

What is the median number of points per game? Explain your reasoning.

TIP

> Remember that the median score is the line inside of the box.

Short-Response Question

2 The following stem-and-leaf plot shows the hourly wages of 20 workers. The workers are performing similar tasks but, because of differences in the amount of work experience and skill, are paid at different rates.

Hourly Wages of 20 Workers

10	5 6 7 7
11	3 4 4 6 6 9
12	0 0 5 5 7
13	0 0 0 1 3

Key
10 | 5 = $10.50

What is the percentage of workers who are paid less than $12 per hour?

TIP

Count the number of workers who are paid less than $12 an hour. This number should be the numerator and the denominator is 20, the total number of workers. Divide 20 into 10 to get a decimal.

3 Karen wants to determine the favorite musical band of the students in her high school. Which sample should she use?

A. a random sample of the students in the environmental club

B. a random sample of the students on the cheerleading squad

C. a random sample of the students in the library during fifth period

D. a random sample of students on the official school roster

TIP

Choose the sample that would not be biased in any way.

Chapter 4 Practice Problems

Complete each of the following practice problems. Check your answers at the end of this chapter. Be sure to read the answer explanations!

1 Mr. Sassy constructed a diagram to illustrate the number of seniors enrolled in honors courses.

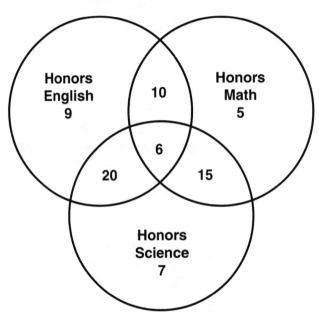

How many seniors are enrolled in Honors Math?

A. 15

B. 20

C. 36

D. 45

Open-Response Question

2   The scatter plot below shows high- and low-temperature information for Dallas, Texas, for six consecutive months.

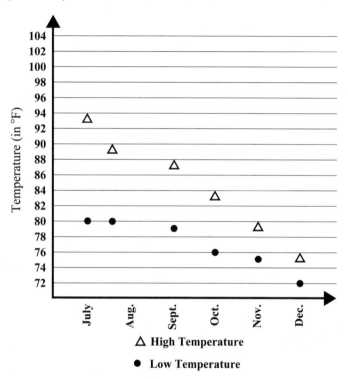

△ High Temperature

● Low Temperature

a. Determine the mean high temperature. Show or explain how you made your determination.

b. Draw a conclusion from the information plotted in the scatter plot.

c. A friend would like to vacation in Dallas, but she doesn't like very high temperatures. During what month would you suggest that your friend go on her vacation? Explain your answer.

3 The box-and-whisker plot below shows the number of points students in Ms. Garcia's class scored on a test.

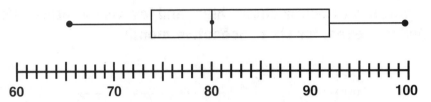

Which statement can be made about the data, using the graph above?

A. The mean score on the test was 92.

B. The mean score on the test was 100.

C. The highest score on the test was 92.

D. The highest score on the test was 100.

Short-Response Question

4 The ages of the residents on the third floor in West Side Retirement Home are plotted on the stem-and-leaf plot below.

Resident Age

6	0, 0, 3, 4, 5, 7, 8
7	0, 1, 4, 7, 9
8	0, 2, 3, 4, 6
9	0, 0, 0, 1, 2

Key
6 │ 0 = 60 years

What percentage of residents are 80 years old or older?

5   

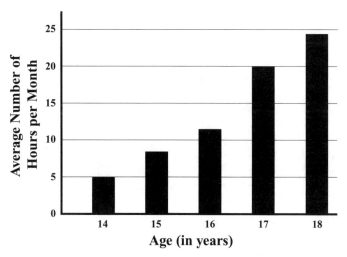

How many hours per month on average did the 17-year-olds work?

A.   8

B.   12

C.   20

D.   24

6   **How many more hours did the 18-year-olds work than the 14-year-olds?**

A.   5 hours

B.   19 hours

C.   24 hours

D.   29 hours

7 The results of a poll asking "What kind of music is your favorite?" are shown in the circle graph below.

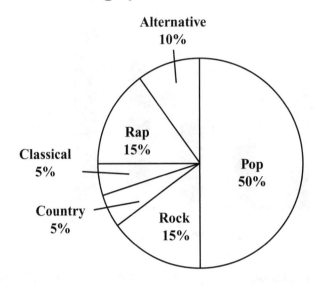

What percentage of people selected either alternative, pop, or classical as their favorite kind of music?

A. 25%

B. 60%

C. 65%

D. 75%

Open-Response Question

8 In a certain town, there are five realty companies, namely A, B, C, D, and E. Use the graph from Company A's advertisement pictured below to answer the question.

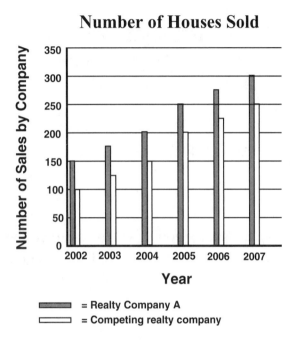

Number of Houses Sold

= Realty Company A
= Competing realty company

a. Identify one way in which the graph "Number of Houses Sold" is misleading. Explain how it is misleading.

b. Identify one way in which the graph is not misleading.

c. What could be done to correct the misleading feature that you identified in part (a)?

Chapter 4 Answer Explanations

Let's Review 8: Graphs

1. C

Four students received five extra-credit points and two students received seven extra-credit points. If you subtract these two numbers, you get two.

2. C

A circle graph is the best way to display percentages.

3. B

To correctly answer this question, you have to subtract the number of students with a dog for a pet, which is $23 + 8 + 2 + 3 = 36$, from the number of students with a cat for a pet, which is $33 + 8 + 2 + 3 = 46$. Then $46 - 36 = 10$.

4. A

The number in the part where all three circles overlap is two.

Let's Review 9: Plots

1. 32

The median score is the line inside the box. In this case, it's 32.

2. 50%

To answer this question, put the number of workers earning less than \$12, which is 10, over the total number of workers: 20. The fraction $\frac{10}{20}$ can be reduced to $\frac{1}{2}$, which is equal to 50%.

3. D

Answer choice A might be biased, since students in the environmental club may share similar interests; the same is true for answer choice B. Answer choice C might be biased, because students of the same age might be in the library during fifth period. Answer choice D is the best answer.

Chapter 4 Practice Problems

1. C

This information can be found in the circle marked "Honors Math": $5 + 10 + 6 + 15 = 36$ students are enrolled in Honors Math.

2. Sample answer:

   a. To determine the mean high temperature, add the high temperatures: 94, 90, 88, 84, 80, and 76. The mean of these temperatures is about 85.

   b. You can conclude that both the high and low temperatures drop steadily from July through to December.

   c. If the friend doesn't like high temperatures, she should plan her vacation to Dallas during the month of December, when it is the coolest.

3. D

The only conclusion supported by the plot is that the highest score was 100.

4. 45%

To determine the answer to this question, count the number of residents who are age 80 or older. There are 10. The total number of residents is 22. If you reduce $\frac{10}{22}$ to $\frac{5}{11}$ and divide 11 into 5, you get .45.

5. C

To find the answer to this question, you have to look at the average number of hours the 17-year-olds worked. It's 20.

6. B

The 18-year-olds worked 24 hours and the 14-year-olds worked 5 hours. If you subtract 5 from 24, you get 19.

7. C

If you add the percentages for pop (50), alternative (10), and classical (5), you get 65.

8.

   a. The bar graph is misleading because it doesn't say what realty company or companies it is comparing to Realty Company A. Realty Company A could have chosen a realty company with very poor sales to compare itself to.

   b. It is not misleading in the way it presents Realty Company A's sales. This seems accurate.

   c. The graph could be redrawn to compare the number of houses sold by each of the five realty companies.

Chapter 5

Geometry, Part 1

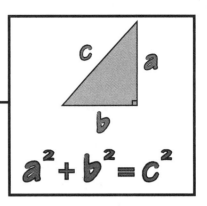

Standards

10.G.1 Identify figures using properties of sides, angles, and diagonals. Identify the figures' type(s) of symmetry.

10.G.2 Draw congruent and similar figures using a compass, straightedge, protractor, and other tools such as computer software. Make conjectures about methods of construction. Justify the conjectures by logical arguments.

10.G.3 Recognize and solve problems involving angles formed by transversals of coplanar lines. Identify and determine the measure of central and inscribed angles and their associated minor and major arcs. Recognize and solve problems associated with radii, chords, and arcs within or on the same circle.

10.G.4 Apply congruence and similarity correspondences (e.g., $\triangle ABC \approx \triangle XYZ$) and properties of the figures to find missing parts of geometric figures and provide logical justification.

10.G.5 Solve simple triangle problems using the triangle angle sum property and/or the Pythagorean theorem.

10.G.6 Use the properties of special triangles (e.g., isosceles, equilateral, 30°-60°-90°, 45°-45°-90°) to solve problems.

In this chapter, you'll learn some properties for plane (flat) figures. And you'll learn how to tell if figures are congruent and similar. It's important to learn the properties of the figures discussed in this chapter, since you will be asked questions about them. You'll also learn about lines and angles in this chapter and you'll learn how to use the Pythagorean theorem to find the length of a missing side in a triangle.

Congruent Figures

Figures that are **congruent** are exactly the same size and shape. If you place two congruent figures on top of each other, they will fit exactly. The rectangles below are congruent. The sign indicating that figures are congruent is ≈.

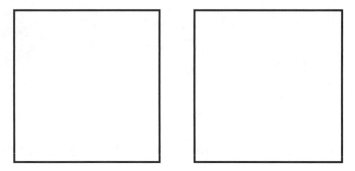

These triangles are also congruent:

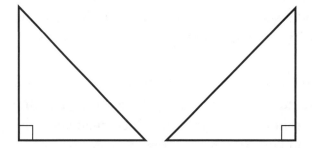

Plane Figures

Some questions on the MCAS will ask you about the properties of plane figures. A **plane figure** is a flat, closed figure. The chart below lists some common plane figures.

PLANE FIGURES

Polygon	A closed figure formed by three or more line segments. A triangle is a polygon. A square is also a polygon. A regular polygon is both equilateral and equiangular.	
Quadrilateral	A polygon with four sides. A square and a rectangle are examples of quadrilaterals.	
Rectangle	Has four right angles and four sides; sides across from each other are parallel and equal in length. The sum of the angles in a rectangle is 360°.	
Square	Has four right angles and four sides equal in length. The sum of the angles in a square is 360°.	
Rhombus	Has four equal sides, may or may not have right angles. The sum of the angles in a rhombus is 360°.	
Parallelogram	Has four sides and two pair of opposite, parallel sides, may or may not have right angles. The sum of the angles in a parallelogram is 360°.	
Trapezoid	Has four sides and one pair of parallel sides; may or may not have a right angle. The sum of the angles in a trapezoid is 360°.	
Triangle	Has three sides; may or may not have a right angle. The sum of all angles in a triangle is 180°.	

(continued)

PLANE FIGURES (continued)

Circle	Has no sides. The sum of the degrees in a circle is 360°.	
Pentagon	Has five sides that may or may not be equal. The sum of the angles in a pentagon is 540°. In a regular pentagon, each angle measures 108°.	
Hexagon	Has six sides that may or may not be equal. Note that a regular hexagon has equal sides. The sum of the angles in a hexagon is 720°. In a regular hexagon, each angle is 120°.	
Octagon	Has eight sides that may or not be equal. Note that a regular octagon has equal sides. The sum of the angles in an octagon is 1,080°. Each angle in a regular octagon measures 135°.	

Similar Figures

Congruent figures are exactly the same shape and size. You could place one congruent figure on top of another and it would fit perfectly. If figures are congruent, the sign ≈ is used, as in $\triangle KLM \approx \triangle NOP$. **Similar figures** are not congruent. Similar figures have the same shape but not the same size.

If figures are similar, the ~ sign is used, as in $\triangle KLM \sim \triangle NOP$.

If figures are similar, their corresponding sides can be written as a proportion because one figure is an enlargement of the other.

These triangles are similar:

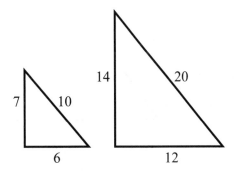

These rectangles are similar:

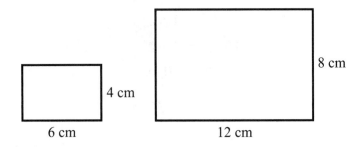

Note that the sides of these rectangles can be written as a proportion that is the same when it is reduced:

$$4 : 6 = 2 : 3$$
$$8 : 12 = 2 : 3$$

On the MCAS, you might be asked to determine the missing side for similar triangles. Read the following question:

Find the missing length (*x*) for the pair of similar triangles below.

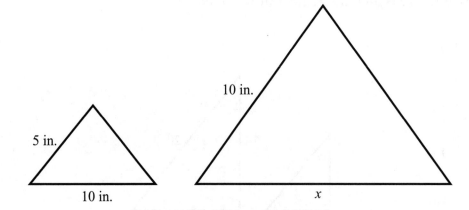

Remember that similar figures have dimensions that can be expressed in the same proportion. The proportion for the first triangle is shown here:

$$5 : 10 = 1 : 2$$
$$10 : x = 1 : 2$$
$$x = 20 \text{ in.}$$

Let's Review 10: Figures and Shapes

Complete each of the following questions. Use the Tip below each question to help you choose the correct answer. When you finish, check your answers with those at the end of Chapter 5.

1 Jamie drew a rectangle that measures 12 inches in width and 24 inches in length. If Jamie enlarges the rectangle so that it's 2 feet wide, how long will the rectangle be?

A. 2 ft

B. 4 ft

C. 6 ft

D. 8 ft

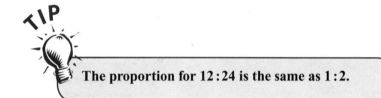

TIP

The proportion for 12 : 24 is the same as 1 : 2.

2 Determine the length of DE. Show your work or provide an explanation to support your answer. Triangles ABC and DEF are similar.

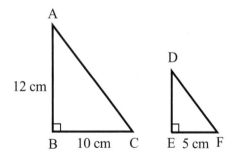

A

12 cm

B   10 cm   C

D

E 5 cm F

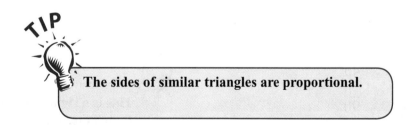

TIP

The sides of similar triangles are proportional.

3   **If two right triangles are similar, find the measure of side *x*.**

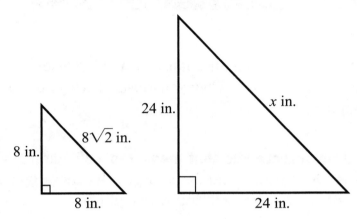

A.   12 in.

B. $24\sqrt{2}$ in.

C.   24 in.

D.   48 in.

Compare a side in the smaller triangle with its corresponding side in the larger triangle.

4   **Look at the triangle below.**

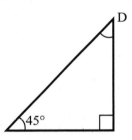

What is the measure of ∠D?

A.   30º

B.   40º

C.   45º

D.   90º

This is a right triangle, so one angle is 90º. Remember that the sum of all angles in a triangle is 180º.

5 Look at the regular pentagon below.

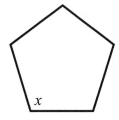

What is the measure of angle *x*?

A. 108°

B. 120°

C. 135°

D. 540°

TIP

If you're unsure of the answer, go back and study the properties of plane figures.

Lines and Angles

A **line segment** is part of a line. It has two **endpoints,** one at each end, to show that it stops and doesn't keep on going.

A **ray** is also part of a line, but unlike a line segment, it keeps on going in one direction. A ray has only one endpoint.

Two rays join together to form an **angle**. The place where they join is called the **vertex.**

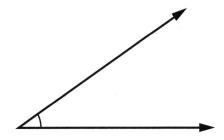

A **line** has an arrow on both ends to show that it keeps going.

Lines that never intersect are called **parallel lines.** Strings on a guitar are parallel, like the lines shown here:

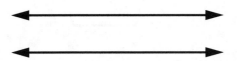

Lines that intersect to form right angles are called **perpendicular lines.** Perpendicular lines form right angles. The place where the lines intersect is called the **point of intersection**.

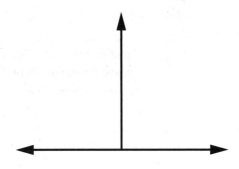

Angles

Angles can be classified by their degree (also called their measure). The following shows some angles that you need to know for the MCAS.

Acute angle

Less than 90 degrees.

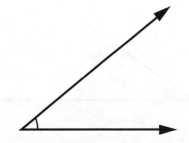

Right angle

Exactly 90 degrees.

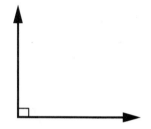

Obtuse angle

Greater than 90 degrees and less than 180 degrees.

Straight angle

Exactly 180 degrees.

Reflex angle

Greater than 180 degrees.

Angle Relationships

The sign ∠ stands for the word "angle." **Adjacent angles** are angles that share a side.

In the illustration below, ∠ADC and ∠CDB are adjacent angles.

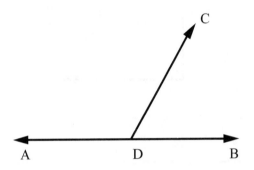

Two angles that add up to 90° are called **complementary angles**. Two angles that add up to 180° are called **supplementary angles**. The angles above are supplementary. ∠ADC measures 125° and ∠CDB measures 55°.

Vertical angles, angles across from each other, are always equal. In the illustration below, angles a and c are equal and angles b and d are equal.

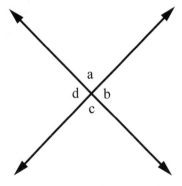

Sometimes parallel lines are intersected by a line. This intersecting line is called a **transversal** and it creates eight angles, four of which are acute and four of which are obtuse. Look at these parallel lines intersected by a transversal:

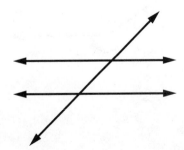

When parallel lines are intersected by a transversal, the four acute angles are always equal and the four obtuse angles are always equal. Each acute and obtuse angle pair forms a supplementary angle, whose sum is 180°.

Let's Review 11: Lines and Angles

Complete each of the following questions. Use the Tip below each question to help you choose the correct answer. When you finish, check your answers with those at the end of Chapter 5.

1   **What is the sum of the measures of complementary angles?**

A.   80°

B.   90°

C.   120°

D.   180°

If you're not sure of the total measure of complementary angles, reread this information in the previous section of this chapter.

2   **Two streets in Josh's neighborhood run next to each other in the same direction but do not intersect. These streets are an example of what kind of lines?**

A.   perpendicular

B.   adjacent

C.   supplementary

D.   parallel

Try to remember the name for two lines that do not intersect.

Triangles

A **triangle** is a plane figure with three sides. Each of the three points on a triangle is called a vertex. You read earlier in this chapter that the sum of the angles in a triangle is 180°.

Equilateral triangle

Has three equal sides and three equal angles; each angle is 60°.

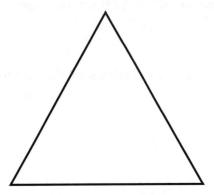

Isosceles triangle

Has two equal sides and two equal angles. For example, an isosceles triangle might have angles measuring 80°-50°-50°.

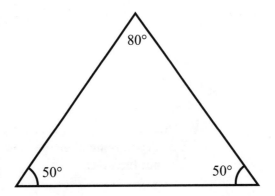

Scalene triangle

Has no equal sides and no equal angles.

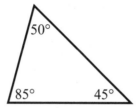

Right triangle

Has one right angle. The side opposite the right angle is called the **hypotenuse.** The other two sides are called **legs**. The legs do not have to be equal.

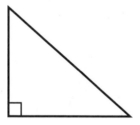

Pythagorean Theorem

The **Pythagorean theorem** is a formula used to find the length of one side of a right triangle when you know the length of the other two. The formula is $a^2 + b^2 = c^2$, where a and b are the lengths of the legs and c is the length of the hypotenuse.

Look at the triangle below.

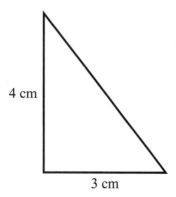

To find the length of the hypotenuse in this triangle, substitute 3 cm and 4 cm into the formula $a^2 + b^2 = c^2$:

$$3^2 + 4^2 = c^2$$
$$9 + 16 = 25$$

Then find the square root of 25. $\sqrt{25} = 5$.

The length of the hypotenuse is 5 cm.

If you are given the length of side c, the hypotenuse, but are missing the length of either leg a or leg b, you can still use the Pythagorean theorem to find the missing side. Look at this triangle:

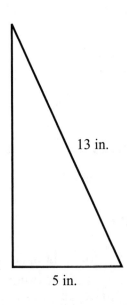

13 in.

5 in.

$$c^2 - a^2 = b^2$$
$$13^2 - 5^2 = b^2$$
$$b^2 = 169 - 25$$
$$b^2 = 144$$
$$b = \sqrt{144} = 12 \text{ in.}$$

Let's Review 12: Triangles

Complete each of the following questions. Use the Tip below each question to help you choose the correct answer. When you finish, check your answers with those at the end of Chapter 5.

1

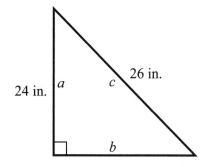

24 in. a c 26 in.

b

In the drawing above, the length of side *a* equals 24 inches. The length of side *c* is 26 inches. Which formula would determine the length of side *b*?

A. $a^2 + c^2 = b^2$

B. $b^2 = a^2 - c^2$

C. $a^2 - b^2 = c^2$

D. $c^2 - a^2 = b^2$

TIP

Remember that you need to put side *b* on one side of the equation.

2   **Which of the following is an equilateral triangle?**

A.

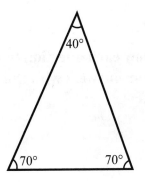

B.

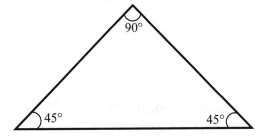

C.

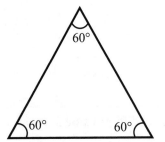

D.

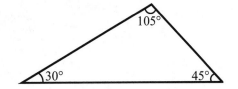

TIP

An equilateral triangle has equal sides.

Short-Response Question

3 **To help a tree grow straight, a landscaper attached a brace and a wire to the tree. He then attached the wire to a stake in the ground.**

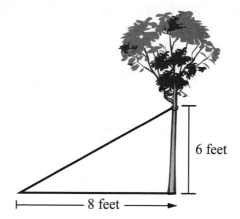

6 feet

8 feet

The brace is 6 feet from the ground and the stake is 8 feet from the base of the tree.

What is the length of the wire?

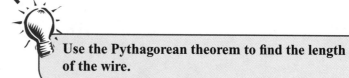

Use the Pythagorean theorem to find the length of the wire.

Chapter 5 Practice Problems

Complete each of the following practice problems. Check your answers at the end of this chapter. Be sure to read the answer explanations!

1   The hexagon below is regular. What is the measure of each of its angles?

A.   90°

B.   108°

C.   120°

D.   180°

2   An irregular octagon has a perimeter of 64. Seven of its sides measure 4, 4, 9, 8, 8, 5, and 10. What is the length of the remaining side?

A.   10

B.   12

C.   14

D.   16

3 What is the measure of ∠DHE?

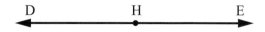

A. 60°

B. 90°

C. 180°

D. 360°

4 Elm Street and Maple Street are parallel to each other. Walnut Street crosses Elm Street and Maple Street. What is the measure of angle *a*?

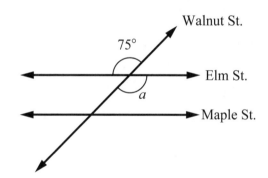

A. 20°

B. 25°

C. 75°

D. 80°

5 Two streets in Terry's neighborhood intersect and form four right angles. These streets are an example of what kind of lines?

A. perpendicular

B. adjacent

C. supplementary

D. parallel

6 Look at the right triangle below. What number is closest to the length of side *c*?

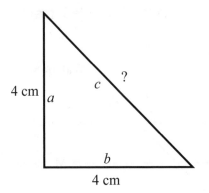

A. 4cm

B. 6 cm

C. 8 cm

D. 16 cm

7 A polygon was drawn on a piece of paper.

● It has two pairs of parallel sides.

● The sum of the measures of its interior angles is 360 degrees.

Which of the following could be the polygon?

A. an equilateral triangle

B. a regular pentagon

C. a regular octagon

D. a parallelogram

8 The measurements, in centimeters (cm) of the sides of a right triangle are shown below.

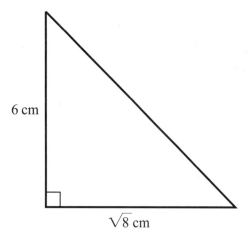

6 cm

$\sqrt{8}$ cm

Which of the following is the length of the hypotenuse of this right triangle?

A. $\sqrt{14}$ cm

B. $\sqrt{28}$ cm

C. $\sqrt{44}$ cm

D. $\sqrt{48}$ cm

Chapter 5 Answer Explanations

Let's Review 10: Figures and Shapes

1. B

The ratio for the smaller triangle is $1:2$. If the enlarged triangle is 2 feet wide, it will be 4 feet long.

2. Sample answer:

The smaller triangle is half the size of the first. You can tell this by looking at the height of the triangle. The larger triangle has a length of 10 cm and the smaller one has a height of 5 cm. Therefore, the ratio is $2:1$. Line DE is the height of the smaller triangle. The larger triangle has a height of 12 cm, so the smaller triangle has a height of 6 cm.

3. B

The ratio for these triangles is $1:3$ and the corresponding side of the smaller triangle is $8\sqrt{2}$, so the correct answer is $24\sqrt{2}$.

4. C

This is a right triangle, so one angle is 90 degrees. One angle of the triangle is labeled 45 degrees, so each of the remaining two angles must also be 45 degrees.

5. A

In a regular pentagon, each angle measures 108 degrees.

Let's Review 11: Lines and Angles

1. B

The sum of complementary angles is 90 degrees.

2. D

Lines that do not intersect are parallel.

Let's Review 12: Triangles

1. D

Since you know the measure of sides a and c, you would subtract the squares of these amounts.

2. C

An equilateral triangle has equal sides and equal angles.

3. Sample answer:

$$6^2 + 8^2 = c^2$$
$$36 + 64 = 100$$
$$\sqrt{100} = 10$$

Chapter 5 Practice Problems

1. C

A regular hexagon has angles that measure 120 degrees.

2. D

To find this answer, add all of the sides of the octagon and subtract from the perimeter.

3. C

The measure of a straight angle is 180 degrees.

4. C

Angle a is congruent to the angle labeled 75°.

5. A

Two lines that intersect to form four right angles are perpendicular.

6. B

If you use the Pythagorean theorem, you'll see that the length of side c is $\sqrt{32}$, which is closest to 6 cm.

7. D

This description matches a parallelogram.

8. C

Using the Pythagorean theorem, and letting $x =$ the length of the hypotenuse, $6^2 + \left(\sqrt{8}\right)^2 = x^2$, $36 + 8 = x^2$, $44 = x^2$, thus $x = \sqrt{44}$.

Chapter 6

Geometry, Part 2

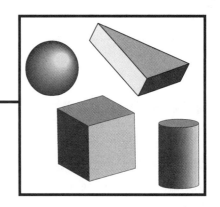

Standards

10.G.7   Using rectangular coordinates, calculate midpoints of segments, slopes of lines and segments, and distances between two points, and apply the results to the solutions of problems.

10.G.8   Find linear equations that represent lines either perpendicular or parallel to a given line and through a point, e.g., by using the "point-slope" form of the equation.

10.G.9   Draw the results, and interpret transformations on figures in the coordinate plane, e.g., translations, reflections, rotations, scale factors, and the results of successive transformations. Apply transformations to the solutions of problems.

10.G.10 Demonstrate the ability to visualize solid objects and recognize their projects and cross sections.

10.G.11 Use vertex-edge graphs to model and solve problems.

10.P.2   Demonstrate an understanding of the relationship between various representations of a line. Determine the line's slope and x- and y-intercepts from its graph or from a linear equation that represents a line. Find a linear equation describing a line from a graph or a geometric description of the line, e.g., by using the "point-slope" or "slope y-intercept" formulas. Explain the significance of a positive, negative, zero, or undefined slope.

In this chapter, you'll learn about some common types of questions on the MCAS about coordinate grids. Some test questions will ask you to move an object on a coordinate grid. Others might ask you to plot an equation on a coordinate grid or to find the slope of a line. You'll also learn about three-dimensional objects in this chapter.

The Coordinate Plane

Some test questions will be about coordinate planes. A **coordinate plane** is a graph with four quadrants, I, II, III, and IV. It has an *x*-axis and a *y*-axis. The *x*-axis is a horizontal line and the *y*-axis is a vertical line. Look at the coordinate plane below. Find the *x*-axis and the *y*-axis, and look at the different quadrants.

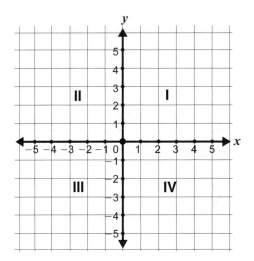

To find coordinates of a point on a coordinate grid, move along the *x*-axis first. If the number of the first coordinate is positive, move to the right. If it's negative, move to the left. Then move along the *y*-axis. If the number is positive, move up. If it's negative, move down. Look at the coordinate grid shown here. Note that the coordinates of point A are (3, 4). Note the coordinates of point B are (3, –4), which is point A reflected over the *x*-axis. Note that the coordinates of point C are (–3, 4), which is point A reflected over the *y*-axis.

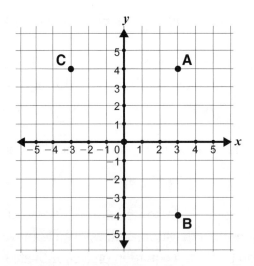

Let's Review 13: The Coordinate Plane

Complete each of the following questions. Use the Tip below each question to help you choose the correct answer. When you finish, check your answers with those at the end of Chapter 6.

1   Give the coordinates of point P on the graph below.

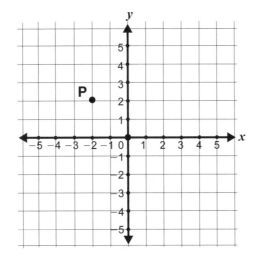

A. $(2, -2)$

B. $(0, 2)$

C. $(-2, 2)$

D. $(1, -2)$

Remember to move along the *x*-axis first. Then move along the *y*-axis.

2 Look at the trapezoid below.

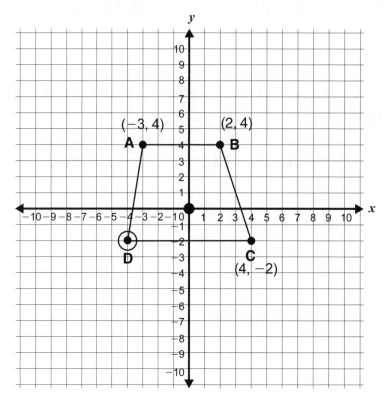

What are the coordinates of point D?

A. $(5, 2)$

B. $(-5, -2)$

C. $(-4, -2)$

D. $(-3, -2)$

TIP

Remember, to find the coordinate of a point, you first move along the *x*-axis and then along the *y*-axis.

3   **Which point on the graph below has the coordinates (−4, −2)?**

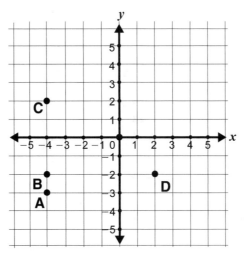

A.   point A

B.   point B

C.   point C

D.   point D

Remember to move along the *x*-axis first.

4 Three of the vertices of a quadrilateral are **(1, 3)**, **(5, 3)**, and **(4, −2)**.

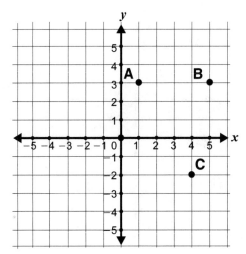

When used as the last vertex, which point would make the quadrilateral a parallelogram?

A. $(1, -2)$

B. $(0, -2)$

C. $(-1, -2)$

D. $(-2, -2)$

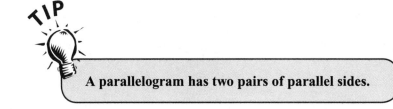

TIP

A parallelogram has two pairs of parallel sides.

Transformations

Some test questions will be about transformations, the movement of figures on a coordinate plane. On the MCAS, you might be asked to choose the correct coordinates of a figure moved in a certain way on the coordinate plane, or you might be asked to move a figure on a coordinate grid. The following are some common transformations.

1. Rotation: When you rotate a figure, you move it around a fixed point, which is called the center of rotation. A rotation can be large or small. A rotation of 180° is called a half-turn. A rotation of 90° is called a quarter turn.

R · Я R · Я

2. Reflection: When a figure is reflected, it is flipped across a line that may or may not be visible. A reflection of a figure is a mirror image.

$$\frac{\text{R}}{\text{Я}}$$

3. Translation: A translation is a "slide." A figure that is translated is moved as if you were sliding it in one direction.

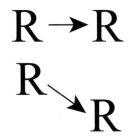

Let's Review 14: Transformations

Complete each of the following questions. When you finish, check your answers with those at the end of Chapter 6.

Open-Response Item

1 **Triangle ABC has vertices with coordinates A (3, 4), B (5, 8), and C (7, 4).**

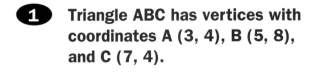

a. Draw and label triangle ABC on the grid provided.

b. Draw the triangle A′B′C′ by translating each vertex of the triangle two units to the left and two units up. Appropriately label the triangle A′B′C′.

c. Draw the triangle A″B″C″ by reflecting triangle A′B′C′ across the *x*-axis.

2 Moving a geometric figure around a fixed point is transformation by

A. inversion.

B. reflection.

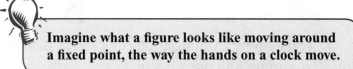

C. rotation.

D. translation.

Imagine what a figure looks like moving around a fixed point, the way the hands on a clock move.

3 On the diagram below, draw the image of triangle ABC reflected over the *x*-axis.

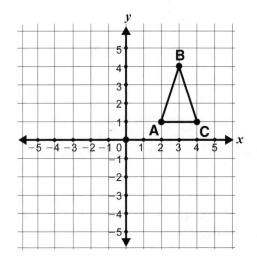

Slope

The **slope** of a line indicates a line's steepness; the greater the slope, the steeper the line. The slope of a line can be positive, negative, or undefined. An **undefined slope** cannot be determined. It might, for example, contain a zero, as in $\frac{3}{0}$.

Lines with a positive slope slant upward from left to right. Lines with a negative slope slant downward from left to right. Lines with 0 slope are horizontal lines. (They are not steep at all!) Horizontal lines have a 0 slope. Vertical lines have an undefined slope.

The **x-intercept** is the pair of coordinates at which a line crosses the *x*-axis, and the **y-intercept** is the pair of coordinates at which a line crosses the *y*-axis.

To determine the slope of a line, use this formula, which is called the **rise over run formula:**

$$\frac{(y_2 - y_1)}{(x_2 - x_1)}$$

Suppose a line has the coordinates listed below. You would use the formula this way:

$$(1, 5) \text{ and } (4, -3)$$
$$x_1, y_1 \qquad x_2, y_2$$
$$\frac{(-3 - 5)}{(4 - 1)}$$

The slope of this line is $-\frac{8}{3}$.

The formula for the equation of any non-vertical line is given by $y = mx + b$, where *m* represents the slope and *b* represents the *y*-intercept. For a vertical line, the equation is given by $x = k$, where *k* is the *x*-intercept.

Vertex-Edge Graphs

You might be asked a test question about a vertex-edge graph. A **vertex-edge graph** has a collection of points, called vertices, and line segments, called edges. Note that in this type of graph, some of the lines are curved. Both curved and straight lines are called edges. Look at the vertex-edge graph below. It has 5 vertices and 8 edges.

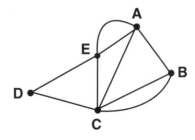

Let's Review 15: Slope and Vertex-Edge Graphs

Complete each of the following questions. Use the Tip below each question to help you choose the correct answer. When you finish, check your answers with those at the end of Chapter 6.

1 Which of the following describes the slope of a line parallel to the *x*-axis?

A. positive slope

B. negative slope

C. zero slope

D. undefined slope

If you don't remember what you learned about slope, reread this section to find the answer.

2 What is the slope of a line that passes through the points (2, 5) and (6, 13)?

A. −2

B. 0

C. 1

D. 2

Use the rise over run formula to find the slope of this line.

3 **A line is shown on the coordinate grid below.**

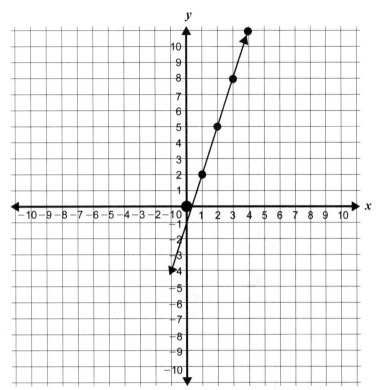

Which of the following best represents an equation of this line?

A. $y = 2x + 1$

B. $y = 3x - 1$

C. $y = \dfrac{1}{3}x + 1$

TIP

Recall that in the form $y = mx + b$, m is the slope (rise divided by run) and b is the y-intercept.

D. $y = \dfrac{1}{2}x - 1$

4 Look at the vertex-edge graph below. How many edges are there?

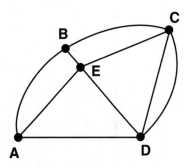

A. 4

B. 6

C. 9

D. 10

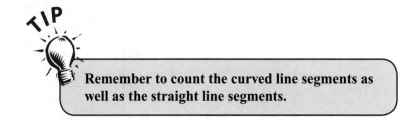

TIP

Remember to count the curved line segments as well as the straight line segments.

Three-Dimensional Figures

Three-dimensional figures are different from plane figures in that they have depth. Unlike plane figures, three-dimensional figures are not flat. It is a good idea to learn the properties of the following three-dimensional figures before taking the MCAS.

Rectangular Solid

Box with six rectangular faces; each corner is a right angle.

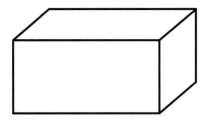

Square Pyramid

Base is a square; has four triangular faces that meet at a vertex.

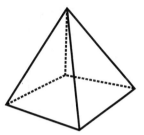

Sphere

A ball where every point on the surface is the same distance from the center.

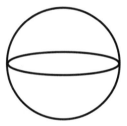

Right Circular Cone

Base is circular; top is shaped like a cone with the vertex directly above the center of the circular base. The distance from the vertex to the circular base is the height.

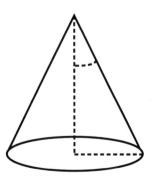

Right Circular Cylinder

Top and bottom are parallel circles; height is the distance from top to bottom.

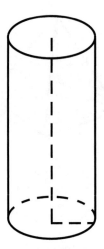

Let's Review 16: Three-Dimensional Figures

Complete each of the following questions. Use the Tip below each question to help you choose the correct answer. When you finish, check your answers with those at the end of Chapter 6.

 Which of the following is a square pyramid?

A.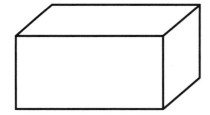

B.

Remember that the base of a square pyramid is a square.

C.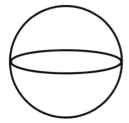

D.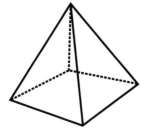

2 **Which of the following has six rectangular faces?**

A.   square pyramid

B.   right circular cylinder

C.   cube

D.   rectangular solid

Reread the previous section, three-dimensional figures, if you're not sure.

Chapter 6 Practice Problems

Complete each of the following practice problems. Check your answers at the end of this chapter. Be sure to read the answer explanations!

1   The vertices of a quadrilateral are $(-8, -3)$, $(-5, -3)$, $(-3, -5)$, and $(-10, -5)$.

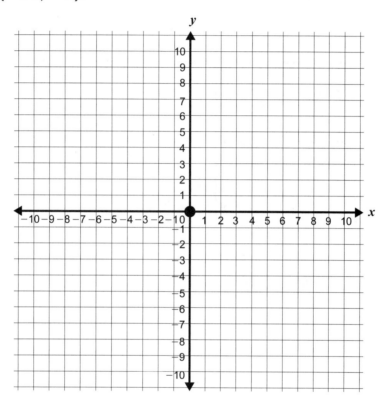

Which describes this quadrilateral?

A.   parallelogram

B.   rectangle

C.   rhombus

D.   trapezoid

2 Cheryl is designing a wallpaper border. She is translating rectangle ABCD to create rectangle A′B′C′D′.

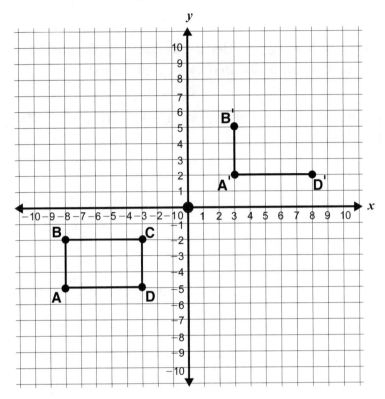

What will be the coordinates of C′?

A. (5, 8)

B. (8, 5)

C. (−5, −3)

D. (−3, −5)

3 Which is the slope of a line that passes through the points (2, 4) and (−7, 10)?

A. $\dfrac{2}{3}$

B. $-\dfrac{2}{3}$

C. 2

D. −2

4 An art student is making geometric designs for a special project. She plots the coordinates of the vertices of a rectangle on a grid. The first three coordinates are (3, 2), (3, 5), and (8, 5). What are the coordinates of the fourth vertex?

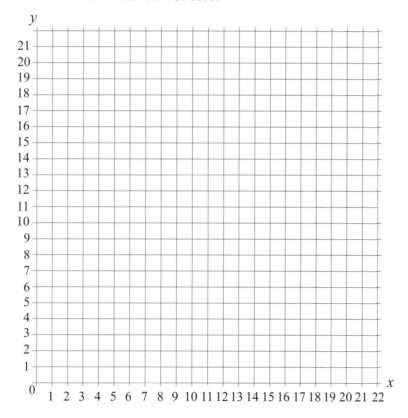

A. (2, 8)

B. (3, 8)

C. (8, 2)

D. (8, 3)

Open-Response Question

5   Four points are connected with line segments, as shown on the coordinate plane below.

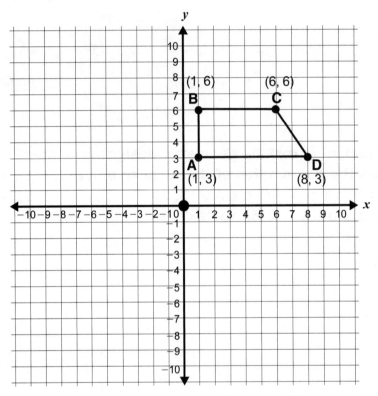

a. Find the slope of each side.

b. Determine if the shape is a trapezoid. Show your work or provide an explanation to your answer.

Short-Response Question

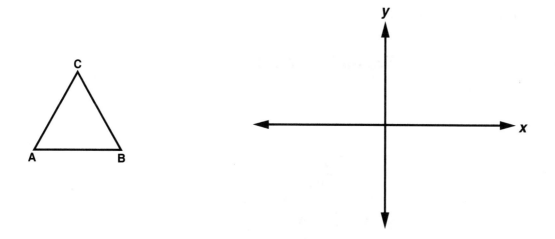

6 Copy equilateral triangle ABC onto the *x*- and *y*-axes shown above, with AB on the *x*-axis and C on the *y*-axis. Then reflect the triangle over the *x*-axis.

Chapter 6 Answer Explanations

Let's Review 13: The Coordinate Plane

1. C
To get to point P, you have to move to -2 on the *x*-axis and 2 on the *y*-axis.

2. C
The coordinates of point D are $(-4, -2)$.

3. B
If you move to -4 on the *x*-axis and -2 on the *y*-axis, you will get to point B.

4. B

In this parallelogram AB = CD and $\overline{AB} \parallel \overline{CD}$. Since AB = 4, just count 4 blocks to the left of point C, which is (0, −2).

Let's Review 14: Transformations

1.

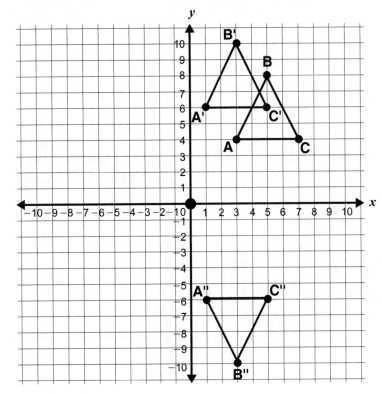

2. C

If you weren't sure of this answer, you could figure it out by using the process of elimination. It is not a reflection or mirror image. Inversion was not discussed, so you can eliminate this answer choice. It's not a translation, which is a slide. It's rotation.

3.

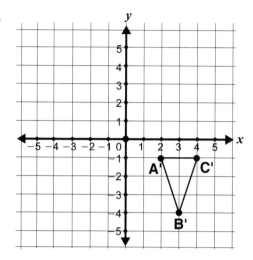

Let's Review 15: Slope and Vertex-Edge Graphs

1. C

Flat horizontal lines have a slope of zero.

2. D

When you substitute the coordinates into the formula, you'll see that the slope is 2.

3. B

Note that the line crosses the *y*-axis at (0, −1). So, only answer choices B and D are possible. Next, look what happens when you move from (1, 2) to (2, 5). The *y* value increases by 3 while the *x* value increases by 1. This implies that the slope is $\frac{3}{1} = 3$. So answer choice B is correct.

4. C

There are four edges around the top of the figure and five edges (including two straight and three curved) around the bottom of the figure.

Let's Review 16: Three-Dimensional Figures

1. D
A square pyramid has a square base and triangular faces.

2. D
A rectangular solid has six rectangular faces.

Chapter 6 Practice Problems

1. D
This figure is a trapezoid, a shape with one pair of parallel sides.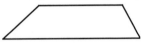

2. B
The coordinates of C′ are (8, 5). Note that point A (–8, –5) is moved to A′ (3, 2), which represents an increase of 11 units in the *x* direction and 7 units in the *y* direction. Likewise, point C′ will be (–3 + 11, –2 + 7) = (8, 5).

3. B
If you substitute these coordinates into the formula to find the slope, you'll see that the slope is $-\frac{2}{3}$. $\frac{(10-4)}{(-7-2)} = \frac{6}{-9} = -\frac{2}{3}$.

4. C
A rectangle has two pairs of equal sides. Therefore, the correct coordinates are (8, 2).

5.   Sample answer:

a.   Using the slope formula, we determine that the slopes of $\overline{AB}$, $\overline{BC}$, $\overline{CD}$, and $\overline{AD}$ are: undefined, 0, $-\dfrac{3}{2}$, and 0, respectively.

b.   The figure is a trapezoid because $\overline{BC}$ and $\overline{AD}$ both have a slope of 0, so they both have parallel sides.

6.

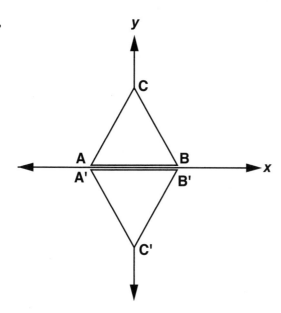

Chapter 7

Measurement

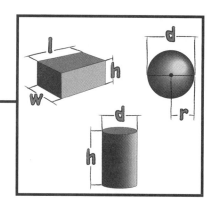

Standards

10.M.1   Calculate perimeter, circumference, and area of common geometric figures such as parallelograms, trapezoids, circles, and triangles.

10.M.2   Given the formula, find the lateral area, surface area, and volume of prisms, pyramids, spheres, cylinders, and cones, e.g., find the volume of a sphere with a specified surface area.

10.M.4   Describe the effects of approximate error in measurement and rounding on measures and on computed values from measurements.

In this chapter, you will learn how to answer questions on the MCAS about measurement. You will build on what you learned in the two chapters on Geometry. You'll work with plane figures again, but this time you'll focus on measurement and learn how to find the perimeter and area of these figures. You'll also learn how to find the volume and surface area of three-dimensional figures.

Many of these test questions will involve real-life situations similar to those you might encounter at home, at work, or at school.

Perimeter

Perimeter is the distance around a plane figure. Remember that a **plane figure** is a flat, closed figure. Perimeter is commonly measured in inches, feet, centimeters, and meters.

The perimeter of a circle is called the **circumference**.

Find the perimeter of the hexagon shown below.

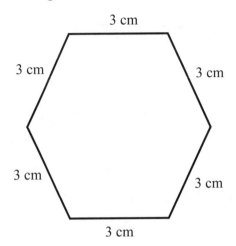

To find the perimeter, add all of the sides: 3 + 3 + 3 + 3 + 3 + 3 = 18, or 6 × 3 = 18 cm.

You just learned that the perimeter of a circle is called the **circumference**. Use this formula to find a circle's circumference: C = π × diameter. Look at this circle:

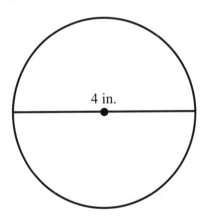

To find the circumference, multiply π (3.14) × the diameter, 4. Then round to the nearest whole number. The circumference is 13 inches.

Sometimes only the radius of a circle is given. When this happens, you have to double the radius, because the diameter is twice the length of the radius. Look at the circle below.

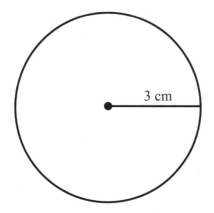

To find the circumference of this circle, first double the radius: $3 \times 2 = 6$. Then plug the numbers into the formula $C = \pi \times$ diameter.

$$C = (3.14) \times 6$$
$$C = 19 \, \text{cm}$$

Central Angles

Angles can be formed inside of a circle when two radii meet and share a vertex. When this happens, the angles are called **central angles**. Look at the circle below. It has two central angles.

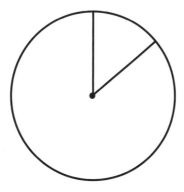

To determine the number of central angles that will fit into a circle, remember that a circle is 360°, and that half of a circle is 180°. For example, in this circle, you can figure out the measure of the missing measurement by subtracting the sum of the measures of the other angles from 360°. The angle in question is 124°.

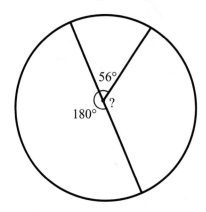

Let's Review 17: Perimeter and Central Angles

Complete each of the following questions. Use the Tip below each question to help you choose the correct answer. When you finish, check your answers with those at the end of Chapter 7.

1 **The perimeters of the two triangles are equal. What is the missing side of the second triangle?**

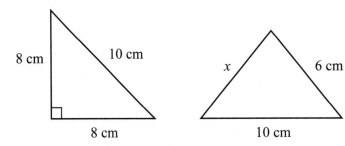

A.   4 cm

B.   6 cm

C.   8 cm

D.   10 cm

To solve this problem, add the sides of the first triangle; then subtract the total of the two sides of the second triangle from the perimeter of the first.

2 A hexagon has a perimeter of 25. Five of its sides are 3, 3, 4, 6, and 6. What is the length of the remaining side?

A. 2

B. 3

C. 4

D. 5

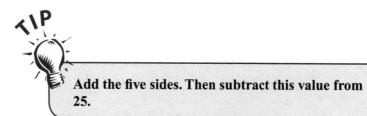

TIP

Add the five sides. Then subtract this value from 25.

3 Look at the clock shown below.

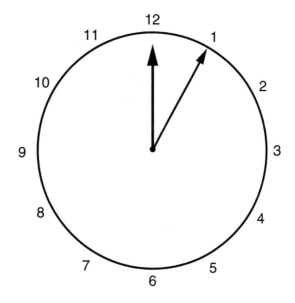

What is the approximate measure of the central angle shown?

A. 12°

B. 15°

C. 30°

D. 36°

TIP

An entire circle measures 360°. This circle is divided into 12 parts.

4   Kelly has an irregularly shaped backyard, as shown below.

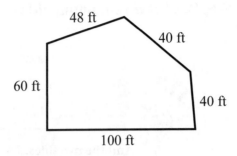

What is the perimeter of his backyard?

A.   188 feet

B.   248 feet

C.   288 feet

D.   388 feet

Add together the sides of Kelly's yard to solve this problem.

Area

For some questions on the MCAS you will have to determine the area of a figure. The following formulas for area are also on your reference sheet:

Rectangle   Area = length × width or $A = lw$, which you may see as base × height or $A = bh$

Triangle   Area = $\frac{1}{2}bh$

Trapezoid   Area = $\frac{1}{2}h(b_1 + b_2)$

Parallelogram   Area = bh

Circle   Area = πr^2

Look at the rectangle below.

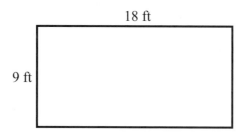

To find the area of this rectangle, plug the length and width into the formula you just learned:

$$A = lw$$
$$A = 18\,\text{ft} \times 9\,\text{ft}$$
$$A = 162\,\text{ft}^2$$

Notice that the area is expressed in square feet.

Look at this circle.

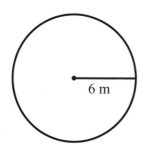

To find the area of this circle, plug the radius into the formula you just learned, which is usually given to you on the test. Use 3.14 for π and round your answer to the nearest whole number:

$$A = (3.14)6^2 \, m^2$$
$$A = (3.14)36 \, m^2$$
$$A = 113 \, m^2$$

Volume

To find the volume or capacity of a rectangular prism or right circular cylinder, use these formulas, which are also on your reference sheet:

Right circular cone $V = \dfrac{1}{3}\pi r^2 h$

Right circular cylinder $V = \pi r^2 h$

Right rectangular prism $V = lwh$ or $V = Bh$ (b = area of a base)

Sphere $V = \dfrac{4}{3}\pi r^3$

Cube $V = s^3$ (s = length of each edge)

Square pyramid $V = \dfrac{1}{3}s^2 h$ (s = length of each side of the base)

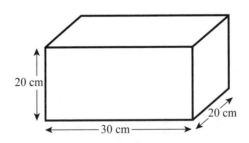

To find the volume of the rectangular prism above, substitute the measurements of the length, width, and height into the formula you just learned:

$$V = 30 \, cm \times 20 \, cm \times 20 \, cm$$

$$V = 12,000 \, cm^3$$

To find the volume of a cylinder, you would use the formula $V = \pi r^2 h$. To find the volume of a cylinder with a radius of 3 and a height of 5, substitute these values into the formula:

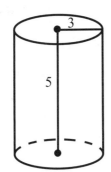

$V = (3.14)3^2 \times 5$

$V = (3.14) \times 9 \times 5$

$V = 141$

Surface Area

The surface area is the total area of the exterior surface of a solid. Lateral surface area does not include the area of the base(s). To find the total surface area for a right circular cylinder and a rectangular prism, use these formulas:

Lateral Surface Area Formulas

right rectangular prism....$LA = 2(hw) + 2(lh)$

right circular cylinder..........$LA = 2\pi rh$

right circular cone ....$LA = 2\pi r\ell$

(ℓ = slant height)

right square pyramid ....$LA = 2s\ell$

(ℓ = slant height)

Total Surface Area Formulas

cube ....$SA = 6s^2$

right rectangular
   prism..................$SA = 2(lw) + 2(hw) + 2(lh)$

sphere ....$SA = 4\pi r^2$

right circular cylinder..........$SA = 2\pi r^2 + 2\pi rh$

right circular cone ....$SA = \pi r^2 + \pi r\ell$

(ℓ = slant height)

right square pyramid ....$SA = s^2 + 2s\ell$

(ℓ = slant height)

To find the surface area of a rectangular solid with a height of 6 cm, a length of 8 cm, and a width of 2 cm, use the second formula shown above:

S.A. $= 2(8 \times 2) = 2 (6 \times 2) = 2 (8 \times 6)$

S.A. $= 2(16) + 2(12) + 2(48)$

S.A. $= 32 + 24 + 96$

S.A. $= 152$ square centimeters

Changing Parameters

Some test questions will ask you what will happen if a parameter in a situation has changed. On the MCAS, these parameters often involve dimension, capacity, cost, and the relationship between two variables.

Suppose you earn $7 an hour at your part-time job and you work 12 hours a week. Before taxes, your paycheck is $84. Now suppose that you have just received a 10% raise. This means that one of the parameters has changed. How much *more* money will you earn now if you work the same number of hours?

To solve this problem, you need to find 10% of $7.

$$\$7 \times .10 = \$0.70$$

Your raise is $0.70 an hour, so your new hourly rate is $7.70. Now multiply $7.70 by 12, the number of hours you work. Your new pay rate is $92.40. Remember that the problem asked you to find how much more money you'll earn, so you need to subtract your old pay rate from your new pay rate:

$$\$92.40 - \$84.00 = \$8.40$$

Let's try one more. Suppose a company manufactures a cylinder-shaped container with a radius of 2 inches and a height of 4 inches. They company would like to make a new container with a radius of 2 inches and a height of 8 inches. How much more liquid will the new container be able to hold?

First, you need to find the volume of the first container. Use the formula to find the volume/capacity of a right circular cylinder on your reference sheet:

$$V = \pi r^2 h$$

Substitute the dimensions of the first container:

$$V = 3.14(2)^2 \times 4$$

$$V = 3.14 \times 4 \times 4$$

$$V = 50.24 \text{ in}^3$$

Now find the volume of the new, larger container:

$$V = 3.14(2)^2 \times 8$$

$$V = 3.14 \times 4 \times 8$$

$$V = 100.5 \text{ in}^3$$

$$100.5 - 50.24 = 50.26 \text{ in}^3$$

The new container will be able to hold 50.26 in.^{3} more than the first container.

Let's Review 18: Area, Volume, Surface Area, and Changing Parameters

Complete each of the following questions. Use the Tip below each question to help you choose the correct answer. When you finish, check your answers with those at the end of Chapter 7.

1   **What is the volume of the box pictured below?**

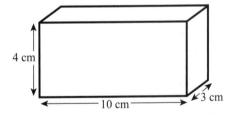

A.   17 cm^3

B.   40 cm^3

C.   70 cm^3

D.   120 cm^3

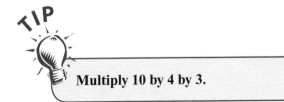

TIP

Multiply 10 by 4 by 3.

2 A rotating sprinkler is used to water a yard. The radius of the area being sprayed is 8 feet. What is the wet area of the yard?

A. 20 square feet

B. 25 square feet

C. 64 square feet

D. 201 square feet

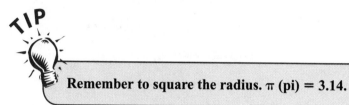

Remember to square the radius. π (pi) = 3.14.

3 The storage tank below has a radius of 8 feet and a height of 10 feet.

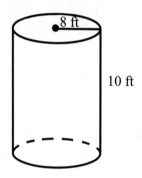

8 ft

10 ft

What is the approximate surface area of the storage tank?

A. 402 square feet

B. 455 square feet

C. 502 square feet

D. 904 square feet

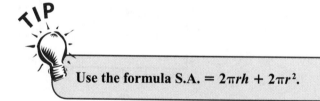

Use the formula S.A. = 2πrh + 2πr².

Short-Response Question

4 The Ramirez family's above-ground pool is shown below. It has a diameter of 12 feet and a height of 5 feet.

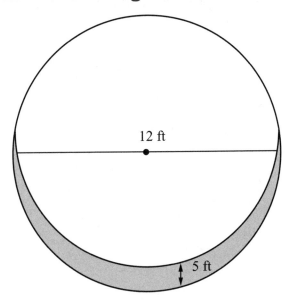

The family would like a new pool. They may increase the height by 1 foot or the diameter by 2 feet.

What is the difference, in cubic feet, in capacity obtained by increasing the diameter 2 feet and the capacity obtained by increasing the height 1 foot? Show your work or explain your answer.

TIP

Determine the volume of the pool with both the increase in the diameter and the increase in the height. Subtract these amounts to find the difference.

Chapter 7 Practice Problems

Complete each of the following practice problems. Check your answers at the end of this chapter. Be sure to read the answer explanations!

1 What is the volume of the box pictured below?

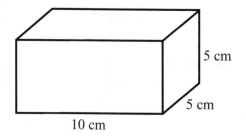

10 cm
5 cm
5 cm

A. 15 cm³

B. 25 cm³

C. 125 cm³

D. 250 cm³

2 Emile wants to wrap the box below with wrapping paper. How many square inches of wrapping paper does he need?

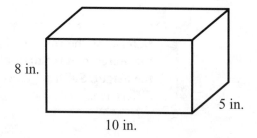

8 in.
10 in.
5 in.

A. 100 square inches

B. 160 square inches

C. 340 square inches

D. 420 square inches

3 Paul wants to close off a rectangular area in his backyard so that he can leave his dog outside. He buys enough fencing to cover 168 square feet.

If the length of the enclosure is 14 feet, what is the width?

A. 12 feet

B. 14 feet

C. 16 feet

D. 18 feet

4 The volume of a cylinder is found by using the formula $V = \pi r^2 h$.

How do the volumes of cylinder A and cylinder B compare?

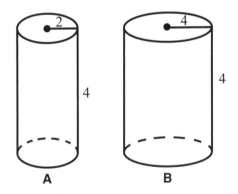

A. The volume of cylinder A is larger.

B. The volume of cylinder B is larger.

C. It is not possible to compare the volumes.

D. The volumes of cylinder A and cylinder B are the same.

5   Rory is 6 feet tall. He casts a shadow that measures 8 feet long at the same time that a tree in the park casts a shadow that is 16 feet long. What is the approximate height of the tree?

A. 12 feet

B. 13.5 feet

C. 15 feet

D. 18.5 feet

6   What is the area of a circle with a radius of 10 cm?

A. 3.14 cm²

B. 31.4 cm²

C. 314 cm²

D. 3,140 cm²

7   Rounding off to the nearest centimeter, what is the volume of the box pictured below?

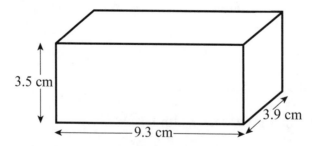

A. 17 cm³

B. 13 cm³

C. 36 cm³

D. 127 cm³

Short-Response Question

8 The diagram below shows the dimensions of a wall that needs to be painted. The door represented by the shaded rectangle is **not** to be painted. The area of the door is 40 square feet.

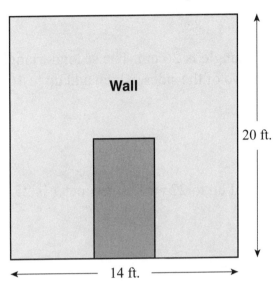

Wall

20 ft.

14 ft.

Determine the area, to the nearest square foot, of the wall that is to be painted. Show your work or provide an explanation for your answer.

Short-Response Question

9 Gretchen earns $10.50 an hour at her summer job. Her employer must pay her "time and a half" ($1\frac{1}{2}$ times her regular hourly earnings) for each hour over 40 hours per week. Her employer withholds 15% of her gross pay for various taxes. The table shows Gretchen's work time for the week.

Gretchen's Hours

Mon	Tues	Wed	Thu	Fri
$9\frac{1}{2}$h	8 h	$8\frac{1}{2}$h	$9\frac{3}{4}$h	$10\frac{1}{4}$h

Determine the amount of Gretchen's paycheck, after taxes are withheld, for the week shown in the table. Show your work or provide an explanation to support your answer.

Chapter 7 Answer Explanations

Let's Review 17: Perimeter and Central Angles

1. D

The perimeter of the first triangle is 26 cm. The second triangle has the same perimeter, but we are only given two of the sides, which add up to 16 cm. Therefore, the last side must be 10 cm.

2. B

The sides of the hexagon add up to 22 and its perimeter is 25. Therefore, the missing side must be 3.

3. C

An entire circle measures 360°. This circle is divided into twelve equal parts and the central angle shown is for one part. Therefore, you need to divide 12 into 360. The answer is 30°.

4. C

To find the perimeter of Kelly's backyard, you have to add the sides: 100 + 40 + 40 + 60 + 48. His yard has a perimeter of 288 feet.

Let's Review 18: Area, Volume, Surface Area, Changing Parameters

1. D

When you multiply 10 by 4 by 3, you get 120.

2. D

To solve this problem, you need to find the area of a circle with a radius of 8 feet. If you substitute this into the formula $A = \pi r^2$, you get approximately 201 ft².

3. D

When you plug in the values for the formula for surface area: S.A. $= 2\pi rh + 2\pi r^2$, it looks like this: $2(3.14) \times 8 \times 10 + 2(3.14) \times 8^2$. When you solve the equation, you get 904 square feet.

4. Sample answer:

If the Ramirez family increases the height of the pool by 1 foot, the volume is calculated like this:

$$V = \pi r^2 h$$
$$V = (3.14)6^2 \times 6 = 678 \text{ ft}^3$$

If the Ramirez family increases the diameter of their pool by 2 feet, the volume is calculated like this:

$$V = (3.14)7^2 \times 5 = 769 \text{ ft}^3$$

The difference in cubic feet is 91 ft³.

Chapter 7 Practice Problems

1. D

Substitute the values in the formula for the volume of rectangular solid: $V = lwh$; $V = 5 \times 5 \times 10.$ $V = 250$ cm³.

2. C

To determine how much wrapping paper Emile needs, substitute the dimensions of the box into the formula to find the surface area of a rectangular square: S.A. $= 2(lw) + 2(hw) + 2(lh)$; S.A. $= 2(8 \times 10) + 2 (10 \times 5) + 2(8 \times 5)$; S.A. $= 2(80) + 2(50) + 2(40)$; S.A. $= 160 + 100 + 80.$ S.A. $= 340$ square inches.

3. A

To find the area of a rectangle, you multiply length by width. This problem gives us the area (168) and the length (14 feet). You need to divide 14 into 168 to get the answer.

4. B

The volume of cylinder A is about 50; the volume of cylinder B is about 200.

5. A

The tree's shadow is twice as long as Roy's shadow. Thus, the tree's height is twice Roy's height.

6. C

When you plug the radius of 10 cm into the formula A = πr^2, you get 314 cm².

7. D

When you multiply the length, width, and height of the box, you get about 127 cm³.

8. Sample answer:

To determine the area of the wall, you need to multiply the length, 20 feet, by the height, 14 feet. The area of the entire wall is 280 square feet. Next, you need to subtract the area of the door. The area of the door is 40 square feet. Therefore, the area of the wall without the door is 240 square feet.

9. Sample answer:

First, you have to add up the hours that Gretchen worked during the week. She worked 46 hours. Next, figure out her pay for 40 hours. 40 × $10.5 = $420. Next, determine her overtime pay: 6 × $15.75 = $94.5. Now add the two amounts: $420 + $94.5 = $514.50. Now determine how much money her employer withholds: $514.50 × 15% = $77.18. Then subtract this from Gretchen's gross pay: $437.32.

Chapter 8

Patterns, Relations, and Algebra, Part 1

$$40x^2$$
$$7(6x)+x$$
$$17-5<20$$

Standards

10.P.3   Add, subtract, and multiply polynomials. Divide polynomials by monomials.

10.P.4   Demonstrate facility in symbolic manipulation of polynomial and rational expressions by rearranging and collecting terms; factoring—(e.g., $a^2 - b^2 = (a + b)(a - b)$, $x^2 + 10x + 21 = (x + 3)(x + 7)$, $5x^4 + 10x^3 - 5x^2 = 5x^2(x^2 + 2x - 1)$; identifying and canceling common factors in rational expressions; and applying the properties of positive integer exponents.

10.P.6   Solve equations and inequalities including those involving absolute value of linear expressions (e.g., $|x - 2| > 5$) and apply to the solution of problems.

10.P.8   Solve everyday problems that can be modeled using systems of linear equations or inequalities. Apply algebraic and graphical methods to the solution. Use technology when appropriate. Include mixture, rate, and work problems.

In this chapter, you'll learn how to answer questions about algebraic expressions and equations. Often you'll be asked to choose the expression or equation that could be used to solve a real-life problem.

You'll see that some expressions and equations have more than one variable, and some of these are displayed in table form. An inequality shows two values that may or may not be equal. You'll learn about inequalities in this chapter.

Expressions

In algebra, letters often stand for numbers that need to be determined. These letters are called **variables**. Keep in mind that any variable, such as x or y, has a 1 before it even though the 1 is not written. In other words, $x = 1x$ and $y = 1y$.

Any number in front of a variable means that the variable will be multiplied by that number. For example, $2x$ means that x will be multiplied by 2.

An algebraic **expression** has a variable, but not an equal sign. Parentheses are often used in algebraic expressions.

Look at this example of an expression:

$$2(x + y)$$

On the MCAS, you will be asked to simplify simple expressions, such as the one shown here:

$$4(3x) + x$$

Begin simplifying this expression by multiplying:

$$12x + x$$

Now simplify it even further by adding. Remember that x is the same as $1x$:

$$13x$$

Other test questions will ask you to choose the correct expression based on a given situation. Read this question:

Mabel bought 12 pencils for p cents each, 6 pens for 9 cents each, 2 erasers for 9 cents each, and 5 sheets of construction paper for p cents each. Which expression would enable Mabel to find out how much money she spent?

To write an expression for this question, you need to add items with the same cost. For example, pencils and construction paper cost p cents each. So, you can add the 12 pencils and 5 sheets of construction paper and multiply them by p, the cost.

$$17(p)$$

You can also add the pens and erasers, since they both cost 9 cents each:

$$8(9)$$

Next you need to write an expression adding the p-cent items and the 9-cent items:

$$17(p) + 8(9)$$

Let's try one more:

Abraham uses the expression $7x + 10.5y$ to determine the amount he earns at a pay rate of seven dollars an hour plus time and a half for overtime. One week he worked 40 hours, plus 2 hours of overtime.

The information in this problem gives you the values to substitute for x and y. Substitute 40 for x, the number of regular hours Abraham worked, and substitute 2 for y, the number of overtime hours Abraham worked. This expression would help Abraham determine how much money he would earn in a week:

$$7(40) + 10.5(2)$$

Equations

An algebraic **equation** is a statement that says two values are equal. You can spot an equation easily because it has an equal sign, which separates the two sides of the equation. Look at this equation:

$$t + 45 = 100$$

In this equation, t is the variable, which represents the unknown quantity. You can solve the equation, and find the value of t, by subtracting 45 from each side of the equation as shown here:

$$t + 45 - 45 = 100 - 45$$
$$t = 55$$

Some questions on the MCAS will ask you to choose the correct equation based on a situation. Read this problem:

Terry's take-home pay is his gross pay minus the $175 his employer deducts each week for taxes. His take-home pay is $500 a week. Write an equation that could be used to find Terry's gross pay.

Terry's gross pay is the unknown variable, or x. This equation could be used to determine his gross pay:

$$x - \$175 = \$500$$

To solve this equation and determine Terry's gross pay, put x, the variable, on a side by itself:

$$x - \$175 + \$175 = \$500 + \$175$$
$$x = \$675$$

Some questions on the MCAS will involve equations with more than one variable. When solving an equation such as $2x + y = 10$, we say that you are solving for y in terms of x. To solve this equation, move everything except y to the opposite side of the equation.

$$2x + y = 10$$
$$y = 10 - 2x$$

Let's try another one. This time solve for b in terms of a.

$$4b = 16a$$
$$b = \frac{16a}{4}$$
$$b = 4a$$

On the MCAS, equations with more than one variable are sometimes written in table form. The table below shows the values of x and y for the equation $3x - y = 0$.

x	y
2	6
3	9
4	12
5	15
6	?

What value makes the equation true when $x = 6$?

If you substitute 6 for x, the equation looks like this: $3(6) - y = 0$. You could probably figure this out in your head. However, the equation can also be solved this way: $18 - y = 0$, or $18 = y$.

Inequalities

An **inequality** links two expressions that may or may not be equal. Inequalities use the following signs:

>	greater than
<	less than
≥	greater than or equal to
≤	less than or equal to
≠	not equal to

Questions about inequalities might also use words like these: *greater than, less than, between, at least,* and *at most.*

Solving an inequality is very similar to solving an equation. Look at this inequality:

$$3y > 21$$

Put y on a side by itself:

$$y > \frac{21}{3}$$
$$y > 7$$

The variable y can be any number that is greater than 7.

Let's Review 19: Expressions, Equations, and Inequalities

Complete each of the following questions. Use the Tip below each question to help you choose the correct answer. When you finish, check your answers with those at the end of Chapter 8.

1   A sporting goods store had 52 sweatshirts at the beginning of a sale. If *y* represents the number of sweatshirts sold during the sale, which expression shows the number of sweatshirts remaining?

A. $y - 52$

B. $52(y)$

C. $52 - y$

D. $y + 52$

TIP

Remember that the store had 52 sweatshirts before the sale, so the amount sold would have to be subtracted from this amount.

2   The number of plain white straws Cara has is shown by the expression 3*x* + 4, with *x* representing her striped straws. If Cara has 10 striped straws, how many plain white straws does she have?

A. 16

B. 30

C. 34

D. 70

TIP

Evaluate the expression like this: 3(10) + 4.

3 Morgan's age is shown by the expression $a + 3$, where a represents Andrea's age. If Andrea is 9, how old is Morgan?

A. 6

B. 9

C. 12

D. 15

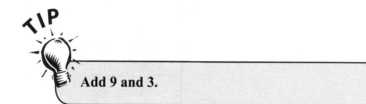

TIP

Add 9 and 3.

4 Solve for y, if $\dfrac{3}{y} = \dfrac{1}{2}$.

A. 2

B. 3

C. 4

D. 6

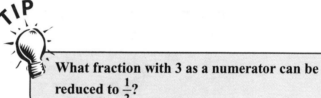

TIP

What fraction with 3 as a numerator can be reduced to $\frac{1}{2}$?

5 A party planner charges a flat fee of $100 to plan a birthday party and an additional $15 per guest. If n = the number of guests and c = total charges, which of the following shows how to determine the total charges?

A. $c = \$100 + \$15n$

B. $c = \$100n + 15$

C. $c = \dfrac{n + 100}{15}$

D. $c = \dfrac{100n}{15}$

TIP

The variable is n, the number of guests.

Short-Response Question

6 The table below shows values for x and y for the equation $x^2 - y = 1$.

x	y
2	3
3	8
4	15
5	24
6	35
7	48
8	?

What value of y makes this equation true when $x = 8$?

Solve for y, then substitute 8 for x.

7 A salesperson's total salary includes a base pay of $800 a month plus 2.5% of the monthly sales. If x = sales per month and y = total salary, which of the following shows how to determine the total salary for any month?

A. $800 = y + 2.5x$

B. $y = \$800 + .025x$

C. $y = \$800x + .025$

D. $y = \$800x + 2.5$

Remember that 2.5% will be added to $800.

8 Given the inequality $3y < 18$, solve for y.

A. $y = 6$

B. $y < 6$

C. $y > 6$

D. $y \leq 6$

TIP

The sign should be the same as the sign in the inequality.

Chapter 8 Practice Problems

Complete each of the following practice problems. Check your answers at the end of this chapter. Be sure to read the answer explanations!

1 A real estate agent is paid each time he sells a house. He is paid a flat fee of $1,000 and 4% of the sale price of the house. If s = sale price of the house and c = total payment, which of the following equations shows how to determine the real estate agent's payment?

A. $c = \$1,000 + .004s$

B. $c = \$1,000 - .004s$

C. $c = \$1,000 + .04s$

D. $c = \$1,000 - .04s$

2 What is the value of x if $3x + 3 = 12$?

A. 2

B. 3

C. 4

D. 9

3 The number of cats at an animal shelter is shown by the expression $2y - 5$, with y representing the number of dogs. If the shelter has 125 dogs, how many cats does it have?

A. 130

B. 245

C. 250

D. 255

Short-Response Question

4 The table below shows values of x and y for the equation $x^3 = y$.

x	y
2	8
3	27
4	?
5	125
6	216
7	343

What is the value of y when $x = 4$?

5 Mrs. Cartez needs to buy wrapping paper to wrap around two rectangular-shaped presents like the one shown below.

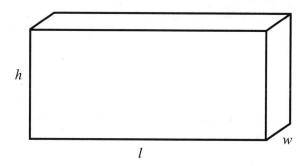

Which expression represents the minimum amount of wrapping paper Mrs. Cartez needs?

A. $l + w + h$

B. $l \times w \times h$

C. $4lw + 4lh + 4wh$

D. $2l + 2w + 2h$

6 At a flea market, a vendor sold 1 handmade quilt and 2 antique plates for less than $100. If q represents the selling price of the quilt and p represents the selling price of one plate, which inequality could be used to show the amount of money the vendor made?

A. $q + p > \$100$

B. $q + 2p < \$100$

C. $q + p \geq \$100$

D. $q + 2p \geq \$100$

Chapter 8 Answer Explanations

Let's Review 19: Expressions, Equations, and Inequalities

1. C

If the store had 52 sweatshirts before the sale and y represents the unknown number of sweatshirts sold during the sale, an expression that could be used to find the number of sweatshirts remaining is $52 - y$.

2. C

To simplify the expression $3x + 4$, put 10 in place of x: $3(10) + 4 = 34$. Cara has 34 plain white straws.

3. C

If Andrea is 9 and Morgan is Andrea's age plus 3, Morgan is 12.

4. D

The fraction $\frac{3}{6}$ can be reduced to $\frac{1}{2}$.

5. A

In this case, the number of guests, n, is the variable, along with the cost. The cost is $100 + $15 times n.

6. 63

If you substitute 8 for x in the equation, $x^2 - y = 1$, you get $64 - y = 1$.

7. B

There are two variables in this equation, the salesperson's total salary and monthly sales. The monthly sales should be multiplied by 2.5% and then added to $800 to determine the total salary for any month.

8. B

To solve this inequality, put y on one side by itself: $3y < 18 = y < \frac{18}{3}$ or $y < 6$.

Chapter 8 MCAS Practice Problems

1. C

The decimal .04 is multiplied by the sale price of the house. Answer choice C is correct.

2. B

To solve the equation $3x + 3 = 12$, put x on one side of the equation: $3x + 3 - 3 = 12 - 3$ or $3x = 9$, $x = 3$.

3. B

To solve this expression, substitute the number of dogs for y in the expression $2y - 5$. $2(125) - 5$ or 245.

4. 64

The number 4 cubed is 64.

5. C

To find the surface area of the rectangular solid, you need to use the expression surface area $= 2lw + 2lh + 2wh$, but in this case, there are two rectangular solids, so you would use the expression 2(surface area) $= 4lw + 4\,lh + 4wh$.

6. B

There are two plates, so you need to put the number two in front of p, the variable representing plates.

Chapter 9

Patterns, Relations, and Algebra, Part 2

1,4,7,10,13...
2,8,32,128...
4,20,100...

Standards

10.P.1   Describe, complete, extend, analyze, generalize, and create a wide variety of patterns, including iterative, recursive (e.g., Fibonacci numbers), linear, quadratic, and exponential functional relationships.

10.P.7   Solve everyday problems that can be modeled using linear, reciprocal, quadratic, or exponential functions. Apply appropriate tabular, graphical, or symbolic methods to the solution. Include compound interest, and direct and inverse variation problems. Use technology when appropriate.

10.P.8   Solve everyday problems that can be modeled using systems of linear equations or inequalities. Apply algebraic and graphical methods to the solution. Use technology when appropriate. Include mixture, rate, and work problems.

Some algebra questions on the MCAS will be about patterns. These patterns may be a repetition of shapes, or numerical patterns in which a like amount is added, subtracted, multiplied, or divided to numbers to form a pattern. You'll learn about the kind of patterns you might see on the MCAS in this chapter.

Suppose you want to plant grass seed in your yard. The clerk at the lawn and garden store tells you that you need to find the area of your yard, meaning you need to know how many square feet of grass seed to buy. You find the area of your yard, but then you see that you need more grass seed than what is available at the store. You'll have to figure out a new area now, the part of your yard that still needs grass seed, because the parameters have changed. You'll learn about changing parameters in this chapter.

Patterns

For some questions on the MCAS, you'll have to determine a pattern to predict what is next in a sequence. This pattern might be in the form of pictures or numbers.

Look at the diagram below.

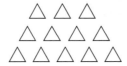

Can you see the pattern in the diagram above? One triangle is taken away for each row that is added. Two triangles will be in the next row, and one triangle will be in the row after that.

Now read this problem:

What is the next number in this geometric sequence?

.04, 0.2, 1.0, 5.0___

You need to determine the pattern before you can find the next number. Look at the numbers. What is the difference between 1.0 and 5.0? It's 4, but this doesn't work for any of the other numbers. However, if you multiply 1.0 by 5 you get 5.0, and this also works for the other numbers. Therefore, the next number in the geometric sequence is 25.0.

Fibonacci Numbers

You might be asked a question with numbers in a pattern referred to as **Fibonacci numbers**. With this pattern, two consecutive numbers are added together to get a third number, as shown here:

1, 1, 2, 3, 5, 8, ...

Let's Review 20: Patterns

Complete each of the following questions. Use the Tip below each question to help you choose the correct answer. When you finish, check your answers with those at the end of Chapter 9.

1 Alice made this design on a wall in her room.

How many flowers will be in the seventh column?

A. 7

B. 14

C. 64

D. 128

Look carefully at the relationship between the first two rows and the third row.

Short-Response Question

2 What is the next number in this geometric sequence?

1.2, 4.2, 14.7, 51.45, _____

TIP

Divide 1.2 into 4.2. Then multiply this number by 51.45.

3 What is the missing number in the pattern?

7.2, 1.8, ___, .1125, .028125

A. 5.4

B. 4.2

C. 2.4

D .45

TIP

Notice that the numbers are decreasing.

4 Look at the letter pattern below.

ABCDABCDABCD

What letter would be in the twentieth place?

A. A

B. B

C. C

D. D

TIP

There is no remainder when you divide 4—the number of letters in the pattern—into 20.

Quadratic Equations

You might be asked a question about a quadratic equation on the MCAS. A quadratic equation is an equation in which the highest exponent of the variable is two. A variable in the equation may be raised to the second degree (squared), but it can't be raised any higher.

The equation below is a quadratic equation:

$$x^2 + 70x + 1,200 = 0$$

You may be asked to choose an equation that is equivalent. This equation is equivalent to the equation above:

$$(x + 30)(x + 40) = 0$$

This can be checked by following the multiplication rules for binomials, namely $(x)(x) + (40)(x) + (30)(x) + (30)(40) = x^2 + 70x + 1,200$

Systems of Equations

Related equations are referred to as a system of equations. On the MCAS, you will be asked to look at a representation and choose the correct system of equations. This system of equations is presented in a matrix. You learned a little bit about matrices in Chapter 2. Look at the system of equations shown here. Remember that you multiply the first member of each row in the first matrix by the top member of the column in the second matrix. Multiply the second members of the rows by the bottom member of the second matrix.

$$\begin{bmatrix} 4 & 2 \\ 5 & 4 \end{bmatrix} \begin{bmatrix} x \\ y \end{bmatrix} = \begin{bmatrix} 20 \\ 14 \end{bmatrix}$$

The multiplication process of matrices is as follows:

$$\begin{bmatrix} 4 & 2 \\ 5 & 4 \end{bmatrix} \begin{bmatrix} x \\ y \end{bmatrix} = \begin{bmatrix} (4)(x) + (2)(y) \\ (5)(x) + (4)(y) \end{bmatrix} = \begin{bmatrix} 4x + 2y \\ 5x + 4y \end{bmatrix}$$

Note that the product is a matrix with 2 rows and 1 column.

These two matrices represent this system of equations:

$$4x + 2y = 20$$
$$5x + 4y = 14$$

Rated Measures

Some questions on the MCAS may ask you to determine the rate at which something happens. For example, you might be asked how far you can travel in 3 hours if you drive at a speed of 55 miles per hour. You will have to multiply or divide to answer these questions. For example:

3 hours $\times$ 55 miles per hour $=$ 165 miles

Let's Review 21: Equations

Complete each of the following questions. Use the Tip below each question to help you choose the correct answer. When you finish, check your answers with those at the end of Chapter 9.

1 If a car travels at 50 miles per hour, about how many miles will it travel in 6 hours?

A. 30 miles

B. 300 miles

C. 3,000 miles

D. 30,000 miles

 TIP
To solve this problem, multiply 50 by 6.

2 $\begin{bmatrix} 2 & 3 \\ 3 & 2 \end{bmatrix} \begin{bmatrix} x \\ y \end{bmatrix} = \begin{matrix} 13 \\ 35 \end{matrix}$

represents which set of equations?

A. $2x + 3y = 13$

  $3x + 2y = 35$

B. $3x + 3y = 13$

  $2x + 3y = 35$

C. $2x - 2y = 13$

  $3x - 2y = 35$

D. $x + 3y = 13$

  $2x + 2y = 35$

 TIP
If you're not sure how to interpret the representation, reread that section in the chapter.

3 Which equation is equivalent to the one below?

$$x^2 + 72x - 5x - 360 = 0$$

A. $(x - 5)(x + 72) = 0$

B. $(x + 6)(x - 12) = 0$

C. $(x - 36)(x - 10) = 0$

D. $(x + 9)(x - 8) = 0$

TIP

Multiply each equation in the answer choices.

4 Ed can type 80 words per minute. How many words can he type in 4 hours?

A. 240 words

B. 320 words

C. 1,920 words

D. 19,200 words

TIP

Multiply to find the answer to this question—and be careful! First convert 4 hours into minutes.

Chapter 9 Practice Problems

Complete each of the following practice problems. Check your answers at the end of this chapter. Be sure to read the answer explanations!

1   The drill team at Karen's school is planning a halftime show. The figure below represents the pattern the team wants to create. The pattern has an innermost ring of 12 members. Each additional ring needs 4 more members than the previous ring.

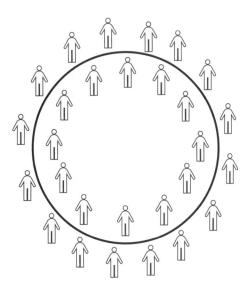

If the halftime show consists of 5 rings formed according to this pattern, what will be the total number of drill team members needed to form all 5 rings?

A.   24

B.   29

C.   48

D.   100

Short-Response Question

2   **What is the next number in this geometric sequence?**

0.5, 1.1, 2.42, 5.324, ____

3   **If the pattern below continued, how many hearts would there be after the fourth row is added?**

A.   8

B.   10

C.   14

D.   22

4 **Which equation is equivalent to the one below?**

$$y^2 - 5y + 20y - 100 = 0$$

A. $(y - 20)(y + 5) = 0$

B. $(y + 20)(y - 5) = 0$

C. $(y + 10)(y - 15) = 0$

D. $(y - 4)(y + 5) = 0$

Chapter 9 Answer Explanations

Let's Review 20: Patterns

1. C
The number of flowers doubles in each column. So the number of flowers in the seventh column would be 64.

2. 180.075
If you divide 1.2 into 4.2, you get 3.5. If you multiply 3.5 by the last number in the series, 51.45, you get 180.075.

3. D
The numbers in this series are divided by 4.

4. D
There are four letters in this series and four goes into 20 evenly, so the pattern completes itself evenly and the last letter is D.

Let's Review 21: Equations

1. B

If you multiply 50 by 6, you get 300.

2. A

The correct set of equations is shown in answer choice A.

3. A

If you multiply $(x - 5)(x + 72) = 0$, you see that it is the same as the quadratic equation.

4. D

To solve this problem, you first need to determine how many minutes are in 4 hours: $60 \times 4 = 240$. Then multiply 80 by this number: $240 \times 80 = 19,200$.

Chapter 9 Practice Problems

1. D

There are two steps needed to solve this problem. First, determine how many drill team members will be in each ring. Then add the number of members in each ring. The first ring has 12 members, the second ring has 16 members, the third ring has 20 members, and so on. The five numbers to be added are 12, 16, 20, 24, 28.

2. 11.7128

1.1 divided by 0.5 is 2.2. Thus, each successive number in this sequence is found by multiplying by 2.2. $(5.324)(2.2) = 11.7128$.

3. D

There are 18 hearts in the diagram now. There is one fewer heart in each row added on. Since there are 5 hearts in the third row, there will be 4 in the fourth row. There will be 22 hearts altogether.

4. B

If you multiply the equation in answer choice B, you'll see that it is equivalent to the quadratic equation.

MCAS Mathematics
Practice Test

Directions: This Practice Test contains 42 questions.

Mark your answers in the answer sheet section at the end of the test.

AREA FORMULAS

square .... $A = s^2$

rectangle...................... $A = bh$

parallelogram .... $A = bh$

triangle .... $A = \frac{1}{2}bh$

trapezoid...................... $A = \frac{1}{2}h(b_1 + b_2)$

circle........................... $A = \pi r^2$

LATERAL SURFACE AREA FORMULAS

right rectangular prism...........$LA = 2(hw) + 2(lh)$

right rectangular prism....... $LA = 2\pi rh$

right circular cone .... $LA = 2\pi r\ell$

$(\ell = \text{slant height})$

right square pyramid .... $LA = 2s\ell$

$(\ell = \text{slant height})$

TOTAL SURFACE AREA FORMULAS

cube .... $SA = 6s^2$

right rectangular prism.........$SA = 2(lw) + 2(hw)$
$+ 2(lh)$

sphere .... $SA = 4\pi r^2$

right circular cylinder.........$SA = 2\pi r^2 + 2\pi rh$

right circular cone .... $SA = \pi r^2 + \pi r\ell$

$(\ell = \text{slant height})$

right square pyramid ....$SA = s^2 + 2s\ell$

$(\ell = \text{slant height})$

CIRCLE FORMULAS

$C = 2\pi r$

$A = \pi r^2$

VOLUME FORMULAS

cube .... $V = s^3$

$(s = \text{length of an edge})$

right rectangular prism ......$V = lwh$

OR

$V = Bh$

$(B = \text{area of a base})$

sphere .... $V = \frac{4}{3}\pi r^3$

right circular cylinder.........$V = \pi r^2 h$

right circular cone .... $V = \frac{1}{3}\pi r^2 h$

right square pyramid ....$V = \frac{1}{3}s^2 h$

SPECIAL RIGHT TRIANGLES

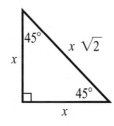 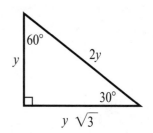

Practice Test 1
Mathematics
SESSION 1

You may use your reference sheet during this session.
*You may **not** use a calculator during this session.*

DIRECTIONS

This session contains fourteen multiple-choice questions, four short-answer questions, and three open-response questions. Mark your answers to these questions on your answer sheet.

1 Which of the following is the largest?

A. 3^4

B. 2^5

C. 5^3

D. 6^2

2 Which of the following is equivalent to the expression below?

$$x + 7x + 3y - y$$

A. $8x + 2y$

B. $10xy$

C. $7x + 3y$

D. $7x + 4y$

3 In a card game, Andy scored 22, 5, 22, 13, 12, 24, 24, 9, 20, and 19 points. What is the **mean** number of points Andy scored?

A. 17

B. 18

C. 19

D. 24

4 How many real numbers are between 1.52 and 1.54?

A. none

B. 100

C. 1000

D. more than 1000

5 A rotating sprinkler is used to water a yard. The radius of the area being sprayed is 12 feet. What is the wet area of the yard?

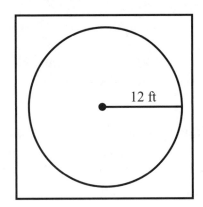

A. 38 square feet

B. 120 square feet

C. 144 square feet

D. 452 square feet

6 What is the value of x in the equation below?

$$9x - 5 = (5)(5) + 15$$

A. 4

B. 5

C. 7

D. 40

7 Which is the slope of a line that passes through the points $(2, 2)$ and $(-3, -3)$?

A. 1

B. -1

C. 2

D. -2

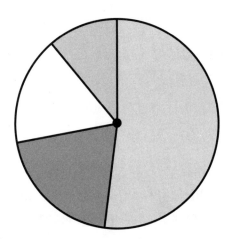

8 Which set of percents would best fit the pie graph shown above?

A. 55%, 25%, 13%, 7%

B. 50%, 18%, 12%, 20%

C. 52%, 20%, 17%, 11%

D. 48%, 22%, 20%, 10%

9 Which of the following must be true for the trapezoid shown below?

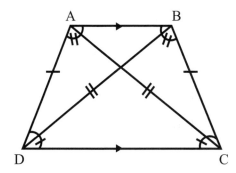

A. Lines AB and CD are parallel.

B. Line BC and AD are perpendicular.

C. ∠A and ∠C are congruent.

D. ∠A and ∠D are congruent.

10 A polygon was drawn on a piece of paper.

- Each of its interior angles is a right angle.
- Opposite sides are congruent.
- Adjacent sides are not congruent

The polygon **must** be which of the following?

A. a rectangle

B. an equilateral triangle

C. a trapezoid

D. a rhombus

11 Kristen has a bag of 40 jelly beans. Five of these jelly beans are pink, 3 are blue, 10 are yellow, 2 are orange, 10 are green, 8 are black, and 2 are white. If Kristen reaches in without looking, what is the probability that she will pull out a white jelly bean?

A. $\dfrac{1}{40}$

B. $\dfrac{1}{20}$

C. $\dfrac{1}{5}$

D. $\dfrac{1}{4}$

12 Find the probability of spinning "blue" on the spinner below.

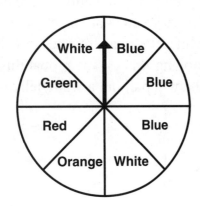

A. $\dfrac{1}{8}$

B. $\dfrac{3}{8}$

C. $\dfrac{1}{4}$

D. $\dfrac{1}{2}$

13 What is the volume of a cube that has an edge of 4 inches?

A.   12 in³

B.   16 in³

C.   64 in³

D.   96 in³

14 What is the next number in this geometric sequence?

9.6, 4.8, 2.4, 1.2 ____

A.   0

B.   0.3

C.   0.5

D.   0.6

Questions 15 and 16 are short-answer questions. Write your answer to these questions in the boxes provided on your answer sheet. Do not write your answers in this test section. You may do your figuring on the answer sheet.

15 The slope of the line containing the points (15,7) and (3,y) is $\frac{3}{4}$. What is the value of y?

16 The following stem-and-leaf plot shows the average temperature in Tampa each day during the month of November.

Stem	Leaf
6	4, 6, 7, 8, 8, 9, 9, 9, 9
7	0, 0, 0, 1, 1, 2, 2, 4, 4, 4, 4, 5, 5, 6, 7, 9
8	0, 0, 1, 1, 1

What percent of days had an average temperature above 70°?

Question 17 is an open-response question.

- **BE SURE TO ANSWER ALL PARTS OF THE QUESTION.**
- **Show all your work (diagrams, tables, or computations) on your answer sheet.**
- **If you do the work in your head, explain in writing how you did the work.**

Write your answer to question 17 in the space provided on your answer sheet.

17 The table below can be used to graph coordinates of x and y.

x	y

a. Fill in the table with the values of x and y in the equation $y = 2x - 1$.

b. Graph the equation in the coordinate grid.

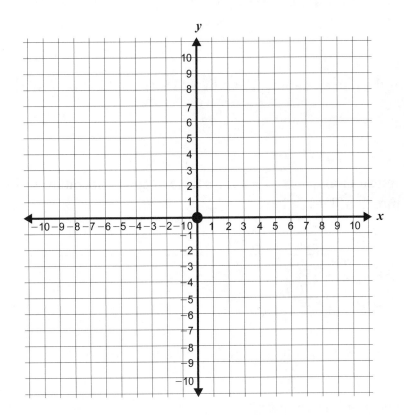

c. Using the coordinates of two of the points you used in the table, determine the slope of the line you graphed.

d. Determine the slope of a parallel line. Explain your answer.

Questions 18 and 19 are short-answer questions. Write your answers to these questions in the boxes provided on your answer sheet. Do not write your answers in this test booklet. You may do your figuring on the answer sheet.

18 A sphere has a surface area of 113.04 square meters. What is the radius of the sphere? Show all your work. Use 3.14 as the value of π.

19 What is the value of the expression below?

$$(6 - 2)^3 + 6 \div 10$$

Question 20 and 21 are open-response questions.

- **BE SURE TO ANSWER AND LABEL ALL PARTS OF EACH QUESTION.**
- Show all your work (diagrams, tables, or computations) on your answer sheet.
- If you do the work in your head, explain in writing how you did the work.

Write your answer to question 20 in the space provided on your answer sheet.

20   Look at the triangles shown below.

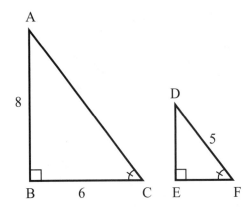

a. Determine the length of the hypotenuse in △ABC.

b. Explain in geometric terms why △ABC is similar to △DEF.

c. Find the length of line ED and EF in △DEF.

Write your answer to question 21 in the space provided on your answer sheet.

21   The prices of tickets to a school musical are shown on the sign below.

Seaside Musical Gala
Adults $15
Children $10

a. During the first night, 300 adult tickets were sold. The total income from ticket sales was $5,600. How many children's tickets were sold during the first night? Show or explain how you got your answer.

For parts (b), (c), and (d), define x and y as follows:

- x = the number of adult tickets sold
- y = the number of children's tickets sold

b. During the second night, a total of 700 tickets were sold. Write an equation that expresses the total number of tickets sold during the second night in terms of x and y.

c. The total income from ticket sales on the second night was $10,000. Write an equation in terms of x and y that expresses the total income during the second night from the sale of adult tickets at $15 each and children's tickets at $10 each.

d. Using your two equations from parts (b) and (c) as a system of equations, solve for x and y. Show or explain how you got your answer.

Practice Test 1
Mathematics
SESSION 2

You may use your reference sheet during this session.
You may use a calculator during this session.

DIRECTIONS
This session contains eighteen multiple-choice questions and three open-ended questions. Mark your answers to these questions on your answer sheet.

22 What is the area of a circle with a radius of 8 centimeters?

   A. 25 cm²

   B. 50 cm²

   C. 201 cm²

   D. 402 cm²

23 **Number of Students in Emma's School**

Grade	Number of Students
1	20
2	28
3	27
4	22
5	25
6	26
7	27

Use the data in the chart to determine the **median**.

   A. 20

   B. 22

   C. 25

   D. 26

24 There are 20 straws in a box; some are red and some are blue. The probability of reaching into the box and pulling out a red straw is $\frac{3}{5}$. How many blue straws are in the box?

   A. 3

   B. 6

   C. 8

   D. 12

25 An irregular hexagon has a perimeter of 36 inches. Five of its sides are 3, 4, 6, 8, and 12. What is the length of the remaining side?

   A. 3 inches

   B. 4 inches

   C. 6 inches

   D. 8 inches

26 What is the value of the expression below?

$$2(4) + 4^2$$

A. 16

B. 22

C. 24

D. 28

27 Find the missing length (x) of the pair of similar figures shown below.

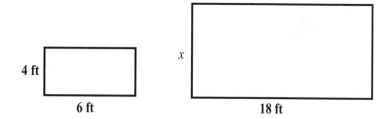

A. 12 feet

B. 20 feet

C. 27 feet

D. 30 feet

28 Which of the following properties of real numbers is demonstrated by the equation below?

$$a(b + c) = a(b) + a(c)$$

A. distributive property

B. inverse property of addition

C. commutative property of addition

D. associative property of addition

29 Which of the following is equivalent to the expression below?

$$5(x - 7y) - 7(x + 5y)$$

A. $-2x - 70y$

B. $-72xy$

C. $-2x$

D. $2x + 70y$

**College Enrollment in Introductory Courses
at Anytown University**

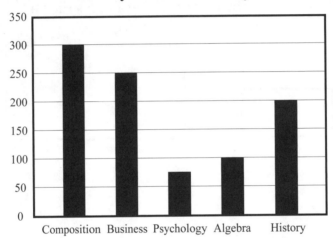

30 The number of students enrolled in introductory courses at a university is shown on the graph above. How many more students are enrolled in composition than in psychology?

A. 200

B. 225

C. 250

D. 375

Question 31 is an open-response question.

- **BE SURE TO ANSWER AND LABEL ALL PARTS OF THE QUESTION.**
- **Show all your work (diagrams, tables, or computations) on your answer sheet.**
- **If you do the work in your head, explain in writing how you did the work.**

Write your answer to question 31 in the space provided on your answer sheet.

31   Use the diagram in the box to answer the question.

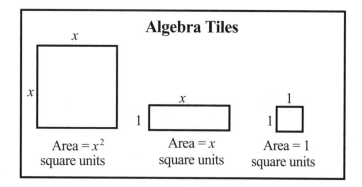

In the figures below, algebra tiles are arranged in a rectangular array to represent each expression.

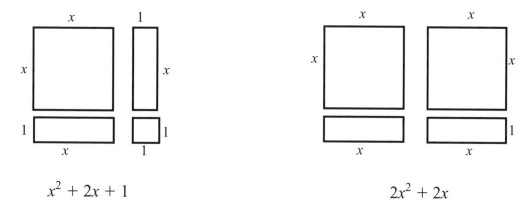

$$x^2 + 2x + 1 \qquad\qquad\qquad 2x^2 + 2x$$

Show how to use algebra tiles to represent each expression below in a rectangular array. If an expression cannot be represented, explain why not.

a. $3x^2 + 3x + 2$

b. $2x^2 + 6x + 4$

Mark your answers to multiple-choice questions on your answer sheet.

32 If the **mean** number of people who visited a museum over 5 days is 250, what is the total attendance during the 5 days?

A. 750

B. 1,000

C. 1,250

D. 2,500

33 To the nearest integer, find the missing length (*x*) of the pair of similar figures shown below.

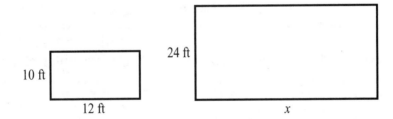

A. 12 feet

B. 20 feet

C. 29 feet

D. 27 feet

34 Emily works in a small crafts store where the cash register does not compute the sales tax. If the sales tax is 6%, what amount should Emily add to a purchase of $12.00?

- A. $0.60
- B. $0.72
- C. $6.00
- D. $7.20

35 Which of the following is the largest?

- A. $2\sqrt{15.5}$
- B. $15.5\sqrt{2}$
- C. $3\sqrt{14.6}$
- D. $14.6\sqrt{3}$

36 If one out of 5 people exercise each day, how many people can be expected to exercise daily in a city of 25,000 people?

- A. 500
- B. 1,500
- C. 2,500
- D. 5,000

Question 37 is a short-answer question.

37 In the figure below, the shaded area is a planar cross section of a rectangular solid.

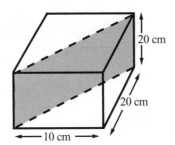

What is the area, in square centimeters, of the cross section?

38 The triangle ABC has vertices at the coordinates $(-5, 1)$, $(-7, 5)$, and $(-3, 5)$, as shown below.

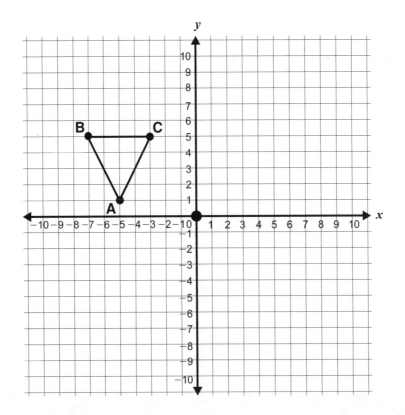

What are the coordinates of the vertices of triangle ABC when it is reflected over the *x*-axis?

    A. $(-7, -5), (-3, -5), (-5, -9)$

    B. $(-5, -1), (-7, -5), (-3, -5)$

    C. $(5, 1), (3, 5), (7, 5)$

    D. $(5, -1), (7, -5), (3, -5)$

39 A square piece of paper, each side six inches long, is folded diagonally on the dotted line, as shown below. To the nearest inch, how long is the crease made in the fold?

    A. 6 inches

    B. 8 inches

    C. 36 inches

    D. 72 inches

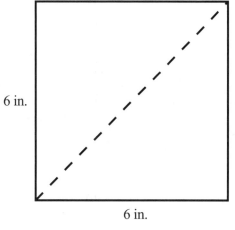

6 in.

6 in.

40 A line is shown on the coordinate grid below.

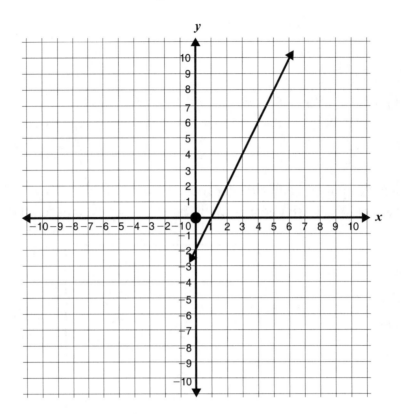

Which of the following best represents an equation of this line?

A. $y = \dfrac{1}{2}x + 2$

B. $y = 2x + 2$

C. $y = 2x - 2$

D. $y = \dfrac{1}{2}x - 2$

Questions 41 and 42 are open-response questions.

- **BE SURE TO ANSWER AND LABEL ALL PARTS OF EACH QUESTION.**
- **Show all your work (diagrams, tables, or computations) in your answer sheet.**
- **If you do the work in your head, explain in writing how you did the work.**

Write your answer to these questions in the space provided on your answer sheet.

41 The total surface area of a cylinder is given by S.A. $= 2\pi rh + 2\pi r^2$, where r is the radius and h is the height.

a. What is the total surface area of a tin can in which the radius is 3 inches and the height is 10 inches? (Round off to the nearest hundredth)

b. The cost of the material for this can is $0.25 per square inch. What is the total cost in making this can? (Nearest hundredth)

c. If the radius is doubled, by what percent will the total cost increase? (Nearest whole number percent)

42 From 1980 to 1990, the population of Kate's town decreased from 150,000 to 75,000.

a. What is the **percent of decrease** in the population over this 10-year period?

b. If the population decreased by the same percent in the next 10-year period, from 1990 to 2000, what was the population in the year 2000?

c. Assume the population continues to decrease at the same rate as it did from 1980 to 1990. Write an expression that would represent the population of Kate's town in the year 2010.

Massachusetts MCAS Grade 10 Mathematics
Practice Test 1
Reporting Categories and Standards

Item No.	Reporting Category	Standard
1	Number Sense and Operations	10.N.1
2	Patterns, Relations, and Algebra	10.P.3
3	Data Analysis, Statistics, and Probability	10.D.1
4	Number Sense and Operations	10.N.1
5	Measurement	10.M.1
6	Patterns, Relations, and Algebra	10.P.3
7	Patterns, Relations, and Algebra	10.P.2
8	Number Sense & Operations	10.N.4
9	Geometry	10.G.1
10	Geometry	10.G.1
11	Data Analysis, Statistics, and Probability	10.D.3
12	Data Analysis, Statistics, and Probability	10.D.3
13	Measurement	10.M.2
14	Patterns, Relations, and Algebra	10.P.1
15	Patterns, Relations, and Algebra	10.P.2
16	Data Analysis, Statistics, and Probability	10.D.1
17	Patterns, Relations, and Algebra	10.P.7
18	Measurement	10.M.2
19	Number Sense and Operations	10.N.2
20	Geometry	10.G.4
21	Patterns, Relations, and Algebra	10.P.7
22	Measurement	10.M.1
23	Data Analysis, Statistics, and Probability	10.D.1
24	Data Analysis, Statistics, and Probability	10.D.3
25	Measurement	10.M.2
26	Number Sense and Operations	10.N.2
27	Geometry	10.G.4
28	Number Sense and Operations	10.N.1
29	Patterns, Relations, and Algebra	10.P.4
30	Data Analysis, Statistics, and Probability	10.D.1
31	Patterns, Relations, and Algebra	10.P.4
32	Data Analysis, Statistics, and Probability	10.D.1
33	Geometry	10.G.4
34	Number Sense and Operations	10.N.1
35	Number Sense and Operations	10.N.1
36	Patterns, Relations, and Algebra	10.P.8
37	Measurement	10.M.2
38	Geometry	10.G.9
39	Measurement	10.M.2
40	Data Analysis, Statistics, and Probability	10.D.2
41	Number Sense and Operations	10.N.1
42	Patterns, Relations, and Algebra	10.P.8

Practice Test 1 Answer Key

Session 1

1. (C)

2. (A)

3. (A)

4. (D)

5. (D)

6. (B)

7. (A)

8. (C)

9. (A)

10. (A)

11. (B)

12. (B)

13. (C)

14. (D)

15. −2

16. 60

17. See explanation.

18. 3

19. 64.6

20. See explanation.

21. See explanation.

Session 2

22. (C)

23. (D)

24. (C)

25. (A)

26. (C)

27. (A)

28. (A)

29. (A)

30. (B)

31. See explanation.

32. (C)

33. (C)

34. (B)

35. (D)

36. (D)

37. 447 cm^2

38. (B)

39. (B)

40. (C)

41. See explanation.

42. See explanation.

MCAS Practice Test 1 Answer Explanations

1. C **(Standard Assessed: 10.N.1)**
Answer choice A is 81, answer choice B is 32, answer choice C is 125, and answer choice D is 36.

2. A **(Standard Assessed: 10.P.3)**
Answer choice A correctly combines the like terms.

3. A **(Standard Assessed: 10.D.1)**
To find the mean number of points Andy scored, total his points (170). Then divide by 10. The answer is 17.

4. D **(Standard Assessed: 10.N.1)**
A real number is any number that can be placed on a number line. Real numbers include fractions. Many, many real numbers would fall between these numbers on a number line.

5. D **(Standard Assessed: 10.M.1)**
To solve this problem, you need to find the area of a circle. To do this, use the formula $A = \pi r^2$. The radius is 12, and 12 squared is 144. The number 144 multiplied by 3.14 is about 452.

6. B **(Standard Assessed: 10.P.3)**
Simplify the equation before solving for x: $9x - 5 = 40$. Now put x on one side of the equation: $x = \frac{(40 + 5)}{9}$. The answer is 5.

7. A **(Standard Assessed: 10.P.2)**
To find the slope of the line, use this formula: $\frac{(y_2 - y_1)}{(x_2 - x_1)}$. The answer is $\frac{5}{5}$ or 1.

8. C **(Standard Assessed: 10.N.4)**
You have to estimate this answer. You can tell by looking at this pie graph that one section is a little over 50% and another is about 10%. The other sections are closest to 20% and 17%.

9. A **(Standard Assessed: 10.G.1)**

All trapezoids have exactly one pair of parallel sides. In an isoceles trapezoid, as we have here, the base angles are also congruent. But the base angles in this case are angles C and D, not angles A and D.

10. A **(Standard Assessed: 10.G.1)**

A rectangle has interior angles that are right angles. It also has opposite sides that are congruent, but adjacent sides are not congruent.

11. B **(Standard Assessed: 10.D.3)**

There are 40 jelly beans in the bag, so this is the denominator. Only 2 of them are white, so 2 is the numerator. When you reduce the fraction, the answer is $\frac{1}{20}$.

12. B **(Standard Assessed: 10.D.3)**

There are eight sections of the spinner, so this is the denominator. Three sections are blue, so the probability of spinning blue is $\frac{3}{8}$.

13. C **(Standard Assessed: 10.M.2)**

A cube has the same length, width, and height, so you would substitute 4 into the formula $V = s^3$ which equals 64 cubic inches or in^3.

14. D **(Standard Assessed: 10.P.1)**

Each of the numbers in the sequence is multiplied by .5.

15. **(Standard Assessed: 10.P.2)**

Using the slope definition, $\frac{y-7}{3-15} = \frac{3}{4}$, $4y - 28 = (3)(-12)$, $4y - 28 = -36$, $4y = -8$, $y = -2$.

16. **(Standard Assessed: 10.D.1)**

To solve this problem, put the number of days that the temperature was over 70 degrees, 18, over the total number of days, 31. Then divide 31 into 18, which is approximately .58 = 58%.

17.   **(Standard Assessed: 10.P.7)**

Sample answer:

a.

x	y
1	1
2	3
3	5
4	7

b.

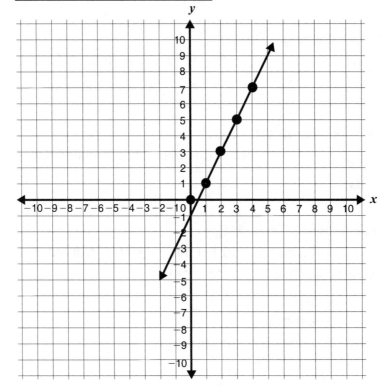

c.   (2, 3) (3, 5) can be substituted into the equation:

$$\frac{y_2 - y_1}{x_2 - x_1}$$

$$\frac{5 - 3}{3 - 2}$$

The slope is 2.

d.   A parallel line will also have the slope of 2, since parallel lines have the same slope.

18.   **(Standard Assessed: 10.M.2)**

Sample answer:

To solve this problem, you need to substitute values into the formula to find the surface area of a sphere:

$$SA = 4\pi r^2$$

$$113.04 = (4)(3.14)r^2$$

$$113.04 = 12.56r^2$$

$$113.04 \,/\, 12.56 = r^2$$

$$9 = r^2$$

$$3 = r$$

19.   **(Standard Assessed: 10.N.2)**

$$(4)^3 + 6 \div 10$$

$$64 + 6 \div 10 = 64 + 0.6 = 64.6$$

20.   **(Standard Assessed: 10.G.4)**

Sample answer:

a. $a^2 + b^2 = c^2$

$8^2 + 6^2 = c^2$

$64 + 36 = 100$

$\sqrt{100} = 10$

$c = 10$

b. m∠A $= 180° - 90° -$ m∠C and m∠D $= 180° - 90° -$ m∠F. Since ∠C ≅ ∠E, ∠A ≅ ∠D. The two triangles are similar because there are 3 pairs of congruent angles.

c. Since triangle DEF is $\frac{1}{2}$ the size of triangle ABC for each corresponding line segment, ED $= \left(\dfrac{1}{2}\right)(8) = 4$ and EF $= \left(\dfrac{1}{2}\right)(6) = 3$.

21. **(Standard Assessed: 10.P.7)**

   a. $300 \times \$15 = \$4,500$

       $\$5,600 - \$4,500 = \$1,100$

       $\$1,100 \div \$10 = 110$, the number of children's tickets sold

   b. $x + y = 700$

   c. $15x + 10y = \$10,000$

   d. $x + y = 700$

       $15x + 10y = 10,000$

       Multiply the first equation by 10 to get $10x + 10y = 7000$. Now subtract the second equation to get $-5x = -3000$, so $x = 600$. Put 600 into the first equation in place of x. Then $600 + y = 700$, so $y = 100$.

22. C **(Standard Assessed: 10.M.1)**

To find the area of a circle, use the formula πr^2. This circle has a radius of 8 cm, so substitute 8 into the formula: $(3.14)64 = 200.9$ or 201 cm.

23. D **(Standard Assessed: 10.D.1)**

The median is the number in the middle. If you put the numbers in order from least to greatest, you'll see that 26 is in the middle.

24. C **(Standard Assessed: 10.D.3)**

Since there are 20 straws in the box, convert $\frac{3}{5}$, the number of red straws, into a fraction with 20 as the denominator: $\frac{12}{20}$. Therefore, eight straws out of the 20 are blue.

25. A **(Standard Assessed: 10.M.2)**

If the perimeter of the hexagon is 36, add the length of the sides you were given: 3, 4, 6, 8, and $12 = 33$. You know that the perimeter is 36, so subtract 33 from 36 to find the missing side.

26. C **(Standard Assessed: 10.N.2)**

$2(4) + 4^2$ or $8 + 16 = 24$.

27. A   (Standard Assessed: 10.G.4)
The measurements of the first rectangle are 4 ft. and 6 ft. Since $18 = 3 \times 6$, the other side would be 3×4 or 12. $\frac{4}{x} = \frac{6}{18}$; $6x = 72$; $x = 12$.

28. A   (Standard Assessed: 10.N.1)
The equation shows the distributive property, which allows you to multiply a sum by multiplying each addend separately and then adding the products.

29. A   (Standard Assessed: 10.P.4)
$5(x - 7y) - 7(x + 5y) = 5x - 35y - 7x - 35y$. Combining like terms, we get $-2x - 70y$.

30. B   (Standard Assessed: 10.D.1)
About 300 students are enrolled in composition, and 75 are enrolled in psychology. If you subtract 75 from 300, the answer is 225.

31.   (Standard Assessed: 10.P.4)
Sample answer.

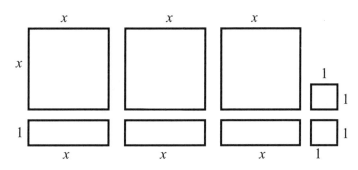

a.   This doesn't work because a rectangle can't be formed.

b.

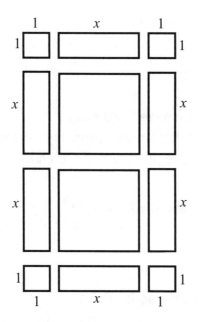

32. C (Standard Assessed: 10.D.1)

To find out how many people attended the museum over five days, multiply the mean by 5.

33. C (Standard Assessed: 10.G.4)

Set up a ratio:

$$\frac{10}{24} = \frac{12}{x}$$

$$24 \times 12 = 288$$

$$10x = 288$$

$$x = 28.8$$

$$28.8 \approx 29$$

34. B (Standard Assessed: 10.N.1)

To find the amount of tax Emily needs to add, multiply $12 by .06. The answer is $0.72.

35. D (Standard Assessed: 10.N.1)

To answer this question, you need to find the value of each of the answer choices. Answer choice A is about 8, answer choice B is about 22, answer choice C is about 11, and answer choice D is about 25. Therefore, answer choice D is the largest.

36. D   (Standard Assessed: 10.P.8)

If one out of 5 people exercise and you want to determine how many people out of 25,000 exercise, divide 5 into 25,000. The answer is 5,000.

37.   (Standard Assessed: 10.M.2)

Sample answer: $a^2 + b^2 = c^2 = 10^2 + 20^2 = c^2$

$$100 + 400 = 500$$

$$\sqrt{500} = 22.36 = c = l$$

$$A_{1234} = lw = (22.36)(20) = 447 \, \text{cm}^2$$

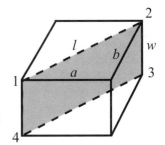

38. B   (Standard Assessed: 10.G.9)

Remember that a reflection is a mirror image. Therefore, when the triangle is reflected over the x-axis, it will look as if it's upside-down.

39. B   (Standard Assessed: 10.M.2)

You have to use the Pythagorean theorem ($a^2 + b^2 = c^2$) to solve this problem. Both side a and side b of the triangle measure 6 inches, so $36 + 36 = 72$. Then you have to find the square root of 72 to get the correct answer: about $8\frac{1}{2}$ inches, which means choice B is the one to pick.

40. C   (Standard Assessed: 10.P.2)

Notice that the y-intercept of this line is $(0, -2)$. By the slope-intercept formula $y = mx + b$, we can narrow down the correct answer choice as either C or D. Next, consider the points $(0, -2)$ and $(1, 0)$ which both lie on this line. The slope value is given by $\dfrac{0 - (-2)}{1 - 0} = 2$. Thus answer choice C is correct.

41.   (Standard Assessed: 10.M.1)

Sample answer:

    a.   Surface area $= (2\pi)(3)(10) + (2\pi)(3^2) = 60\pi + 18\pi \approx 245.04$ square inches.

    b.   Total cost $= (245.04)(\$0.25) = \$61.26$

    c.   The new radius is 6. The new surface area is $(2\pi)(6)(10) + (2\pi)(6^2) = 120\pi + 72\pi \approx 603.19$ square inches. The new total cost is $(603.19)(\$0.25) = \$150.80$. Thus, the percent increase is

$$\left(\frac{\$150.80 - \$61.26}{\$61.26} \right)(100) \approx 146\% \, .$$

42. **(Standard Assessed: 10.P.8)**

Sample answer:

a. $\dfrac{75{,}000}{150{,}000} = \dfrac{3}{6} = \dfrac{1}{2} \times 100\% = 50\%$

b. 37,500

c. $37{,}500 \times .50 = 18{,}750$

Practice Test 1
SESSION 1

1 Ⓐ Ⓑ Ⓒ Ⓓ   **2** Ⓐ Ⓑ Ⓒ Ⓓ   **3** Ⓐ Ⓑ Ⓒ Ⓓ

4 Ⓐ Ⓑ Ⓒ Ⓓ   **5** Ⓐ Ⓑ Ⓒ Ⓓ   **6** Ⓐ Ⓑ Ⓒ Ⓓ

7 Ⓐ Ⓑ Ⓒ Ⓓ   **8** Ⓐ Ⓑ Ⓒ Ⓓ   **9** Ⓐ Ⓑ Ⓒ Ⓓ

10 Ⓐ Ⓑ Ⓒ Ⓓ   **11** Ⓐ Ⓑ Ⓒ Ⓓ   **12** Ⓐ Ⓑ Ⓒ Ⓓ

13 Ⓐ Ⓑ Ⓒ Ⓓ   **14** Ⓐ Ⓑ Ⓒ Ⓓ

15

16

17

x	y

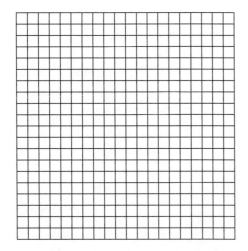

18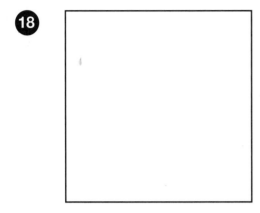

19

20

21

_____

_____

_____

SESSION 2

22 Ⓐ   Ⓑ   Ⓒ   Ⓓ   **23** Ⓐ   Ⓑ   Ⓒ   Ⓓ   **24** Ⓐ   Ⓑ   Ⓒ   Ⓓ

25 Ⓐ   Ⓑ   Ⓒ   Ⓓ   **26** Ⓐ   Ⓑ   Ⓒ   Ⓓ   **27** Ⓐ   Ⓑ   Ⓒ   Ⓓ

28 Ⓐ   Ⓑ   Ⓒ   Ⓓ   **29** Ⓐ   Ⓑ   Ⓒ   Ⓓ   **30** Ⓐ   Ⓑ   Ⓒ   Ⓓ

31 _____

_____

_____

_____

_____

_____

_____

_____

_____

_____

_____

_____

_____

_____

_____

_____

32 Ⓐ Ⓑ Ⓒ Ⓓ   **33** Ⓐ Ⓑ Ⓒ Ⓓ   **34** Ⓐ Ⓑ Ⓒ Ⓓ

35 Ⓐ Ⓑ Ⓒ Ⓓ   **36** Ⓐ Ⓑ Ⓒ Ⓓ

37

38 Ⓐ Ⓑ Ⓒ Ⓓ   **39** Ⓐ Ⓑ Ⓒ Ⓓ   **40** Ⓐ Ⓑ Ⓒ Ⓓ

41

42

MCAS Mathematics
Practice Test

Directions: This Practice Test contains 42 questions.

Mark your answers in the answer sheet section at the end of the test.

AREA FORMULAS

square ....$A = s^2$

rectangle ....$A = bh$

parallelogram ....$A = bh$

triangle ....$A = \dfrac{1}{2}bh$

trapezoid.....................$A = \dfrac{1}{2}h(b_1 + b_2)$

circle............................$A = \pi r^2$

LATERAL SURFACE AREA FORMULAS

right rectangular prism.............$LA = 2(hw) + 2(lh)$

right rectangular prism.........$LA = 2\pi rh$

right circular cone ....$LA = 2\pi r\ell$

(ℓ = slant height)

right square pyramid .... $LA = 2s\ell$

(ℓ = slant height)

TOTAL SURFACE AREA FORMULAS

cube ....$SA = 6s^2$

right rectangular prism ....$SA = 2(lw) + 2(hw)$

$+ 2(lh)$

sphere ....$SA = 4\pi r^2$

right circular cylinder..........$SA = 2\pi r^2 + 2\pi rh$

right circular cone ....$SA = \pi r^2 + \pi r\ell$

(ℓ = slant height)

right square pyramid ....$SA = s^2 + 2s\ell$

(ℓ = slant height)

CIRCLE FORMULAS

$C = 2\pi r$

$A = \pi r^2$

VOLUME FORMULAS

cube ....$V = s^3$

(s = length of an edge)

right rectangular prism ......$V = lwh$

OR

$V = Bh$

(B = area of a base)

sphere ....$V = \dfrac{4}{3}\pi r^3$

right circular cylinder.........$V = \pi r^2 h$

right circular cone ....$V = \dfrac{1}{3}\pi r^2 h$

right square pyramid ....$V = \dfrac{1}{3}s^2 h$

SPECIAL RIGHT TRIANGLES

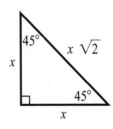
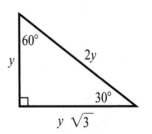

Practice Test 2
Mathematics
SESSION 1

You may use your reference sheet during this session.
You may not use a calculator during this session.

DIRECTIONS

This session contains fourteen multiple-choice questions, four short-answer questions, and three open-response questions. Mark your answers to these questions in the spaces provided on your answer sheet.

1 Frankie has 30 baseball cards at the beginning of the week. If x represents the number of baseball cards Frankie gave to his friend Amy on Tuesday, and y represents the number of baseball cards his mother gave him on Thursday, which expression shows the number of baseball cards Frankie has at the end of the week?

A. $x + 30 - y$

B. $30 - x + y$

C. $30x - y$

D. $30y + x$

2 Find the probability of spinning a "2" on the spinner below.

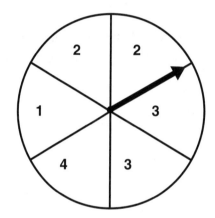

A. $\dfrac{1}{6}$

B. $\dfrac{1}{4}$

C. $\dfrac{1}{3}$

D. $\dfrac{1}{2}$

3 Which number below is the greatest?

  A. $\sqrt{196}$

  B. $3\sqrt{12}$

  C. 2^4

  D. 3^3

4 Jeffrey needs to simplify the following expression on his homework assignment.

$$4(x + 2y) + 2(3x - y) - (x + y)$$

Which of the following expressions is equivalent to the expression above?

  A. $10x - 7y$

  B. $9x + 5y$

  C. $-12x$

  D. $-12x + 7y$

5 Given the inequality $6x < 42$, solve for x.

  A. $x = 7$

  B. $x < 7$

  C. $x > 7$

  D. $x \le 7$

6 Lisa has a bag of 30 marbles. Five of these marbles are white, 3 are blue, 10 are pink, 5 are red, 2 are green, 3 are orange, and 2 are black. If Lisa reaches into the bag and pulls out a marble without looking, what is the probability that she will pull out a red marble?

  A. $\dfrac{1}{30}$

  B. $\dfrac{1}{6}$

  C. $\dfrac{1}{5}$

  D. $\dfrac{1}{4}$

7 Simplify the expression below.

$$\frac{3y}{y^3}$$

  A. $3y^2$

  B. $\dfrac{3y}{y}$

  C. $\dfrac{3}{y^2}$

  D. $\dfrac{3}{y^3}$

8 Teresa plans to set up a lemonade stand at a local fair. She will purchase 250 cans of lemonade for $75 and will charge $2.50 for each can she sells. In addition to the cost of the lemonade, Teresa will need to pay $20 to set up the stand. Which of the following expressions could Teresa use to find out how much money she could make after expenses, for selling x cans of juice?

A. $2.5x - 75 - 20$

B. $x + 2.50 - 75 - 20$

C. $2.50 - 75 - \frac{20}{x}$

D. $2.5x (75 - 20)$

9 Rounding to the nearest inch, what is the volume of the box pictured below?

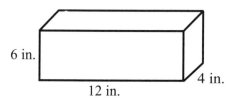

6 in.

12 in.

4 in.

A. 22 in^3

B. 72 in^3

C. 288 in^3

D. 576 in^3

10 Dawn recorded the number of cats adopted at her local animal shelter from January through July.

January	210
February	180
March	212
April	215
May	175
June	195
July	220

What is the **mean** number of cats adopted at the animal shelter?

A. 194

B. 201

C. 208

D. 215

11 Ms. Roberts constructed a diagram to illustrate the number of girls who are enrolled in the athletic program.

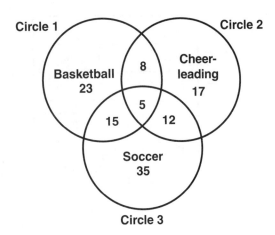

How many girls play both basketball and soccer, but are not cheerleaders?

A. 8

B. 12

C. 15

D. 35

12 Which of the following is closest to the value of $\left(\sqrt{81}\right)^3$?

A. 13.5

B. 27

C. 243

D. 729

13 In the diagram below, triangle ABC is reflected over the *y*-axis and rotated how many degrees about the origin counterclockwise?

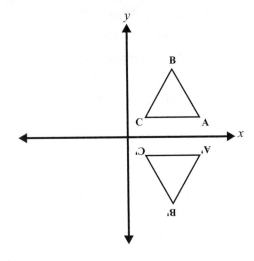

A. 45°

B. 90°

C. 180°

D. 360°

14 The figure shown below is a parallelogram.

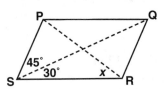

What is the measure of angle *x*?

A. 165°

B. 105°

C. 95°

D. 75°

Questions 15 and 16 are short-answer questions. Write your answer to these questions in the boxes provided on your answer sheet. Do not write your answers in this test section. You may do your figuring on the answer sheet.

15 A high school basketball court that meets the regulation size requirement measures 100 feet long and 50 feet wide.

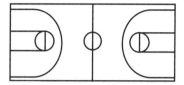

Leo's elementary school uses a basketball court that is similar in size and shape but has a length 20 feet shorter than the high school regulation basketball court. What is the width, in **feet,** of the elementary school basketball court?

16 Rubina purchased 2 pairs of sneakers for $84.99. The sneakers were discounted 20% the next week, and the store manager agreed to give Rubina a refund equal to the amount of the discount. How much would the sneakers have cost if Rubina had waited one week? Show your work in the space below.

Mark your answer to multiple-choice question 17 on your answer sheet.

17 Jose's scores in psychology class are 94, 69, 84, 78, 90, 75, 94, 90, 90, and 95. What is the **mode** of his test scores?

   A. 26

   B. 86

   C. 90

   D. 92

Questions 18 and 19 are short-response questions. Write your answer to these questions in the boxes provided on your answer sheet. You may do your figuring on the answer sheet.

18 Peter bought a mountain bike for $200. When he took the bike home, he noticed a large scratch on one side. The bike shop gave Peter a 35% refund on the bike. How much did Peter pay for the bike after the refund? Show your work.

19 Ashaki joined her school's cross-country team. As part of her training, she is going to increase the distance she runs every week by 2 miles. If she runs 21 miles the first week, how many miles will she run during the eighth week?

Questions 20 and 21 are open-response questions.

- **BE SURE TO ANSWER AND LABEL ALL PARTS OF EACH QUESTION.**
- **Show all your work (diagrams, tables, or computations) on your answer sheet.**
- **If you do the work in your head, explain in writing how you did the work.**

Write your answer to these questions in the space provided on your answer sheet.

20 A contractor uses a 15-foot ladder to reach the roof. The ladder is 9 feet away from the house and forms a right triangle.

a. Find the length of the missing side of this triangle.

b. Determine the area of the triangle.

Write your answer to Question 21 in the space provided on your answer sheet.

21 The average salary for all restaurant workers in a certain area is $275 a week. The weekly salaries of 7 employees at Lobster King are given in the table below.

Employee 1	$200
Employee 2	$225
Employee 3	$240
Employee 4	$240
Employee 5	$280
Employee 6	$375
Employee 7	$400

a. Determine the measures of center of the 7 salaries.

- mean

- median

- mode

- range

b. Specify which of these measures of center the management could use to represent the salaries in an argument against pay increases. Explain your answer.

c. Specify which of these measures of center the labor union could use to represent the salaries in an argument for pay increases. Explain your answer.

Practice Test 2
Mathematics
SESSION 2

You may use your reference sheet during this session.
You may use a calculator during this session.

DIRECTIONS
This session contains seventeen multiple-choice questions and four open-response questions. Mark your answers to these questions in the spaces provided on your answer sheet.

22 In 2006, 131 million people in the United States were employed. Of these, about 13% worked in manufacturing jobs. According to this information, about how many people in the United States were employed in manufacturing jobs?

   A. 13,000,000

   B. 14,500,000

   C. 17,030,000

   D. 34,000,000

23 Melanie uses the expression $8a + 12b$ to determine the amount she earns at a pay rate of 8 dollars an hour plus time and a half for overtime. One week she worked 40 hours, plus 8 hours of overtime. What is her total pay for the week?

   A. 364

   B. 384

   C. 410

   D. 416

24 Daniel is creating a garden in his backyard. He's planting rows of shrubs like the ones shown below. The first row has 1 shrub, the second has 2, the third has 4, and the fourth has 8.

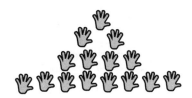

If the pattern continues, how many shrubs would be in the fifth row?

   A. 8

   B. 12

   C. 16

   D. 24

25 Beth scored 18, 18, 15, 18, 18, 24, 21, 20, 24, and 14 points during her first ten basketball games. What is her **mean** score?

A. 10

B. 18

C. 19

D. 21

26 The sphere below is intersected halfway between the top and bottom by a plane parallel to the base of the sphere. The radius is 4 centimeters.

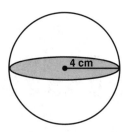

Which number is the **best** approximation of the area of the shaded region?

A. 6 square centimeters

B. 16 square centimeters

C. 50 square centimeters

D. 100 square centimeters

27 A can of soup has a diameter of 3 inches and height of 4 inches.

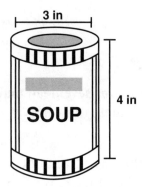

a. What would be the volume of the soup can if the diameter were increased by 1 inch? Figure to the nearest hundredth. Show your work.

b. What would be the volume of the original soup can if the height were increased by 1 inch? Figure to the nearest hundredth. Show your work.

c. The cost to produce a soup can is $0.12 per cubic inch of capacity (volume). What is the difference in cost per can to produce the soup cans in parts (a) and (b)? Figure to the nearest cent. The soup can in which part costs less? Show your work.

28 During the last five years, the population in Frieda's town decreased from 3.6 million to 1.8 million. What was the **percent of decrease** in Frieda's town over the past five years?

A.   25%

B.   35%

C.   50%

D.   60%

29 A clothing store marked all sweaters $\frac{1}{4}$ off the original price for a sale. Alex has a store coupon that is good for an additional discount of 10% off the sale price. She wants to purchase a sweater that was originally priced at $45.00. If she uses her discount coupon, what should be the cost of the sweater before the sales tax is added?

A. $14.55

B. $23.75

C. $30.38

D. $33.75

30 Cecily has a photograph that measures 8 inches wide and 10 inches in length. If Cecily has the photograph enlarged so that it is 24 inches wide, how long will the picture be?

A.   24 inches

B.   30 inches

C.   240 inches

D.   320 inches

Question 31 is an open-response question.

- **BE SURE TO ANSWER AND LABEL ALL PARTS OF THE QUESTION.**
- **Show all your work (diagrams, tables, or computations) on your answer sheet.**
- **If you do the work in your head, explain in writing how you did the work.**

Write your answer to question 31 in the space provided on your answer sheet.

31   Look at the figures below.

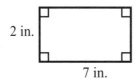

2 in.

7 in.

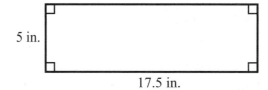

5 in.

17.5 in.

a. Explain in geometric terms why the rectangles are similar.

b. How are the areas related as a ratio, as a reduced fraction, or decimal?

c. If another rectangle with a length of 35 inches is similar to the rectangles shown above, what would be the area of this rectangle?

Mark your answers to multiple-choice questions 32 through 40 in the spaces provided on your answer sheet.

32 Madeline earns $8.25 an hour babysitting her cousins during the 10 weeks of summer vacation. If she averages 12 hours per week, how much does Madeline earn during the summer?

   A. $82.50

   B. $99.00

   C. $825.00

   D. $990.00

33 If a jacket originally cost $75 and is selling at a 20% discount, what is the amount of the discount?

   A. $7.50

   B. $11.25

   C. $15.00

   D. $18.75

34 Determine the volume of a cylinder with a radius of 3 inches and a height of 12 inches. Use 3.14 for π.

$$V = \pi r^2 h$$

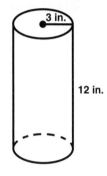

3 in.

12 in.

A. 108 in³

B. 113 in³

C. 226 in³

D. 339 in³

35

$$\begin{bmatrix} 1 & 3 \\ 4 & 3 \end{bmatrix} \begin{bmatrix} x \\ y \end{bmatrix} = \begin{matrix} 9 \\ 24 \end{matrix}$$

represents which system of equations?

A. $y + 3x = 9$
   $4y + 3x = 24$

B. $x + 3y = 9$
   $4x + 3y = 24$

C. $4x + y = 9$
   $3x + 4y = 24$

D. $x + 3y = 9$
   $x + 12y = 24$

36 The population of Gina's town grew from 1.5 million ten years ago to 2.7 million today. What was the **percent of increase** in the population of Gina's town over ten years?

A. 44

B 55

C. 64

D. 80

37 What is the area of a circle with a diameter of 12 centimeters?

A. 36 cm²

B 113 cm²

C. 144 cm²

D. 452 cm²

38 Charlie wants the mean of his 5 English test scores to be at least 85%. His scores on the first 4 tests are 80%, 83%, 90%, and 92%. What is the minimum score Charlie can earn on the 5th test to meet his goal?

A. 75

B. 80

C. 85

D. 90

39 Christa found the following information showing the percentages of students at her school taking part in after-school activities during the past 12 months:

Clubs	75%
Sports	47%
Student Government	8%
Employment	28%
Other	10%

Which type of graph is most appropriate for displaying this information?

A. a line graph

B. a circle graph

C. a scatter plot

D. a bar graph

40 What is the twelfth term in this sequence?

2, 3, 5, 8, 13, ….

A. 192

B. 233

C. 377

D. 610

Questions 41 and 42 are open-response questions.

- **BE SURE TO ANSWER AND LABEL ALL PARTS OF EACH QUESTION.**
- **Show all your work (diagrams, tables, or computations) on your answer sheet.**
- **If you do the work in your head, explain in writing how you did the work.**

Write your answer to these questions in the space provided on your answer sheet.

41 Nora's Catering Service charges a flat fee of $100 per job, plus $10 per guest attending a party.

a. If c represents Nora's total charges, write an equation that expresses c in terms of g, the number of guests attending a party.

b. What is the total charge for Nora's Catering Service for 112 guests? Show all your work.

Villa Mambo's Take-Out, another catering service, charges no flat fee but charges $12 per guest attending a party.

c. If c represents Villa Mambo's total charges, write an equation that expresses c in terms of g, the number of guests attending a party.

d. The Millers plan to have a party catered, and they like Nora's Catering Service and Villa Mambo's Take-Out equally well. If they invite 112 guests, which service would cost more? Show all your work.

Write your answer to question 42 in the space provided on your answer sheet.

 Mr. Rolands is analyzing the scores his 10 honors math students earned on their last test. To make his calculations easier, he reduced each score by 80 points and arrived at the simplified data set shown below.

$$\{0, 0, 1, 2, 4, 4, 4, 6, 7, 8\}$$

a. For each simplified data set, find each of the measures listed below. Show or explain how you got each answer.

- mean

- median

- mode

- range

b. For the set of actual scores on the math test, find each of the measures listed below. Show or explain how you got each answer.

- mean

- median

- mode

- range

Massachusetts MCAS Grade 10 Mathematics Practice Test 2 Reporting Categories and Standards

Item No.	Reporting Category	Standard
1	Patterns, Relations, and Algebra	10.P.4
2	Data Analysis, Statistics, and Probability	10.D.3
3	Number Sense and Operations	10.N.1
4	Patterns, Relations, and Algebra	10.P.4
5	Patterns, Relations, and Algebra	10.P.6
6	Data Analysis, Statistics, and Probability	10.D.3
7	Patterns, Relations, and Algebra	10.P.4
8	Patterns, Relations, and Algebra	10.P.4
9	Measurement	10.M.2
10	Data Analysis, Statistics, and Probability	10.D.1
11	Data Analysis, Statistics, and Probability	10.D.1
12	Number Sense and Operations	10.N.1
13	Geometry	10.G.9
14	Geometry	10.G.1
15	Geometry	10.G.4
16	Number Sense and Operations	10.N.4
17	Data Analysis, Statistics, and Probability	10.D.1
18	Number Sense and Operations	10.N.1
19	Patterns, Relations, and Algebra	10.P.1
20	Geometry	10.G.5
21	Data Analysis, Statistics, and Probability	10.D.1
22	Number Sense and Operations	10.N.4
23	Patterns, Relations, and Algebra	10.P.3
24	Patterns, Relations, and Algebra	10.P.1
25	Data Analysis, Statistics, and Probability	10.D.1
26	Geometry	10.G.10
27	Measurement	10.M.3
28	Patterns, Relations, and Algebra	10.P.8
29	Number Sense and Operations	10.N.1
30	Geometry	10.G.1
31	Geometry	10.G.4
32	Number Sense and Operations	10.N.1
33	Number Sense and Operations	10.N.1
34	Measurement	10.M.2
35	Data Analysis, Statistics, and Probability	10.D.4
36	Patterns, Relations, and Algebra	10.P.8
37	Measurement	10.M.1
38	Patterns, Relations, and Algebra	10.P.1
39	Data Analysis, Statistics, and Probability	10.D.1
40	Patterns, Relations, and Algebra	10.P.1
41	Patterns, Relations, and Algebra	10.P.8
42	Data Analysis, Statistics, and Probability	10.D.2

Practice Test 2 Answer Key

Session 1

1. (B)

2. (C)

3. (D)

4. (B)

5. (B)

6. (B)

7. (C)

8. (A)

9. (C)

10. (B)

11. (C)

12. (D)

13. (C)

14. (B)

15. See explanation.

16. $67.99

17. (C)

18. $130

19. 35

20. See explanation.

21. See explanation.

Session 2

22. (C)

23. (D)

24. (C)

25. (C)

26. (C)

27. See explanation.

28. (C)

29. (C)

30. (B)

31. See explanation.

32. (D)

33. (C)

34. (D)

35. (B)

36. (D)

37. (B)

38. (B)

39. (D)

40. (C)

41. See explanation.

42. See explanation.

MCAS Practice Test 2 Answer Explanations

Session 1

1. B   (Standard Assessed: 10.P.4)

Frankie has 30 baseball cards at the beginning of the week, and he gives x away. So you know the expression will begin with $30 - x$. Then Frankie's mother gives him more baseball cards (y). The entire expression should be $30 - x + y$.

2. C   (Standard Assessed: 10.D.3)

There are six sections on the spinner, so the denominator will be 6. Two of these sections have a 2 on them, so the fraction is $\frac{2}{6}$ or $\frac{1}{3}$.

3. D   (Standard Assessed: 10.N.1)

To answer this question, you need to determine the value of each answer choice. Answer choice A, the square root of 196, is 14. Answer choice B, 3 times the square root of 12, is approximately 10. Answer choice C, 2 raised to the fourth power, is 16. Answer choice D, 3 raised to the third power, is 27. Answer choice D is the correct answer.

4. B   (Standard Assessed: 10.P.4)

The expression $4(x + 2y) + 2(3x - y) - (x + y)$ can be simplified like this:

$4x + 8y + 6x - 2y - x - y$. Then combine like terms to get this result:

$9x + 5y$

5. B   (Standard Assessed: 10.P.6)

The only answer choice that works is $x < 7$, since $(6) \times (7)$ is 42.

6. B   (Standard Assessed: 10.D.3)

The denominator is 30 since this is how many marbles are in the bag. There are 5 red marbles in the bag. This fraction reduces to $\frac{1}{6}$.

7. C   (Standard Assessed: 10.P.4)

You can simplify the expression by canceling out the y in the numerator and one y in the denominator.

8. A   **(Standard Assessed: 10.P.4)**
The variable in this problem is x, the number of cans of lemonade Teresa sells. Teresa will charge $2.50 for each can, so the best expression is $2.5x - 75 - 20$.

9. C   **(Standard Assessed: 10.M.2)**
 If you multiply 12 by 6 and then the answer by 4, you get 288.

10. B   **(Standard Assessed: 10.D.1)**
To find the mean, you need to add all of the numbers (1407) and divide by the number of months (7). $140 \div 7 = 20$.

11. C   **(Standard Assessed: 10.D.1)**
To find out how many girls play basketball and soccer, you need to look in the overlapping portion of circle 1 and circle 3, excluding circle 2.

12. D   **(Standard Assessed: 10.N.1)**
The square root of 81 is 9. The number 9 cubed is 729.

13. C   **(Standard Assessed: 10.G.9)**
The triangle is reflected and then moved completely around the origin. It is moved 180 degrees.

14. B   **(Standard Assessed: 10.G.1)**
In a parallelogram, consecutive angles are supplementary. $m\angle S = 45° + 30° = 75°$. Since $\angle R$ is supplementary to $\angle S$, $m\angle R = 105°$.

15.   **(Standard Assessed: 10.G.4)**
To solve this problem, set up a proportion: $\frac{50}{100} = \frac{x}{80}$. Then solve the proportion:

$$100x = 4,000$$

$$x = \frac{4,000}{100}$$

$$x = 40$$

16. $67.99 per pair (Standard Assessed: 10.N.4)

First determine how much the sneakers are discounted: $84.99 \times .20 = \$17.00$. Then deduct this amount from the original cost of the sneakers.

17. C (Standard Assessed: 10.D.1)

The mode is the number that occurs most; in this case, it is 90.

18. $130 (Standard Assessed: 10.N.1)

To solve this problem, you first need to determine the 35% discount. Multiply .35 by $200 to get $70. Then deduct this amount from $200. Peter's final cost was $130.

19. 35 (Standard Assessed: 10.P.1)

Begin with the number 21, the number of miles Ashaki runs during the first week. Then, add 2 to this number until you get to the eighth week.

20. (Standard Assessed: 10.G.5)

Part A

$$a^2 + b^2 = c^2$$

$$9^2 + b^2 = 15^2$$

$$b^2 = 15^2 - 9^2$$

$$b^2 = 225 - 81$$

$$b^2 = 144$$

$$b = 12 \text{ feet}$$

Part B

$$A = \frac{1}{2}bh$$

$$A = \frac{1}{2}(12 \times 9)$$

$$A = \frac{1}{2}(108)$$

$$A = 54 \text{ square feet}$$

21. **(Standard Assessed: 10.D.1)**

Sample answer: $200 + 225 + 240 + 240 + 280 + 375 + 400 = 1,960$

    a. $1,960 \div 7 = 280$. This is the mean. The mode is 240 and the median is 240. The range is 200.

    b. Management would use the mean in an argument against pay raises.

    c. The labor union could use either the mode or median in an argument for pay raises.

Session 2

22. C **(Standard Assessed: 10.N.4)**

To find the number of people employed in manufacturing jobs, multiply 131,000,000 by .13 or 13%.

23. D **(Standard Assessed: 10.P.3)**

Melanie worked 40 hours, for which she earned $8 an hour. She earned $320 for the 40 hours. Then she worked 8 hours of overtime, for which she is paid $12 an hour. She earned $96 in overtime. If you add $320 + $96, the answer is $416.

24. C **(Standard Assessed: 10.P.1)**

The number of shrubs doubles in each row, so the fifth row would have 16 shrubs.

25. C **(Standard Assessed: 10.D.1)**

When you add Beth's scores, the answer is 190. When you divide this by 10, the number of basketball games, the answer is 19.

26. C **(Standard Assessed: 10.G.10)**

Use the formula $A = \pi r^2$ to solve this problem. $A = (\pi)(4)^2 = 16\pi \approx 50.3$, which is closest to 50.

27. **(Standard Assessed: 10.M.3)**

   a. The dimensions of the soup can become: diameter = 4 inches and height = 4 inches. This means that the new radius becomes 2 inches. Then $V = (\pi)(R^2)(H) = (\pi)(2^2)(4) = 16\pi \approx 50.27$ cubic inches.

   b. The dimensions of the soup can become: diameter = 3 inches and height = 5 inches. This means that the radius remains 1.5 inches. Then $V = (\pi)(R^2)(H) = (\pi)(1.5^2)(5) = 11.25\pi \approx 35.34$ cubic inches.

   c. The difference in volume is 50.27 − 35.34 = 14.93 cubic inches. Thus, the difference in cost is (14.93)($0.12) ≈ $1.79. The soup can in part (b) costs less.

28. C **(Standard Assessed: 10.P.8)**

To find the percent of decrease, first subtract the two numbers: 3.6 − 1.8. Then put the difference over the original number, $\frac{1.8}{3.6}$. Then reduce and multiply by 100%: $\frac{1}{2} \times 100\% = 50\%$.

29. C **(Standard Assessed: 10.N.1)**

First determine the sale price: $\frac{1}{4}$ or .25 percent of $45 = 11.25; the sale price is $45 − $11.25 = $33.75. Then take 10% off this price: $33.75 − $3.37 = $30.38.

30. B **(Standard Assessed: 10.G.1)**

To solve this problem, you can set up a proportion: $\frac{8}{10} = \frac{24}{x}$. The ratio between 8 and 24 is 1:3, so the correct answer is 30.

31. **(Standard Assessed: 10.G.4)**

   a. The two rectangles are similar because their sides are proportional. The second is an enlargement of the first by a factor of 2.5.

   b. The area of the smallest rectangle is 14 inches, whereas the area of the larger rectangle is 87.5 inches. The ratio of the areas is $\frac{14}{87.5}$ inches, which can be reduced to .16 or $\frac{4}{35}$ inches.

c. This rectangle has a length that is twice the length of the larger rectangle on the diagram. Thus, its width would be $(2)(5) = 10$ and its area would be 350 units^2.

32. D   **(Standard Assessed: 10.N.1)**

To determine how much money Madeline earns during the summer, multiply her hourly wage, $8.25, by 12, the number of hours she works per week. Then multiply this amount by 10, the number of weeks she works during the summer.

33. C   **(Standard Assessed: 10.N.1)**

To find the discount, multiply $75 by .20. The discount is $15.

34. D   **(Standard Assessed: 10.M.2)**

To find the volume of the cylinder, use the formula: $V = \pi r^2 h$

$$V = (3.14)3^2 \times 12; V = 339 \text{ in}^3$$

35. B   **(Standard Assessed: 10.P.4)**

Answer choice B shows the correct equation for the representation. Remember that with a matrix, you multiply the first member of each row in the first matrix by the top member of the column in the second matrix. Then multiply the second members of the rows by the bottom member of the second matrix.

36. D   **(Standard Assessed: 10.P.8)**

To find the percent increase in the population of Gina's town, subtract 1.5 from 2.7 to get 1.2. Then divide 1.5 into 1.2.

37. B   **(Standard Assessed: 10.M.1)**

To find the area of the circle, use the radius, 6, which is one-half of 12, the diameter. Then, plug 6 cm into the formula $A = \pi r^2$ to get $\pi(6^2) = 36 \pi \sim 113$

38. B   **(Standard Assessed: 10.P.1)**

Charlie's average on the four tests he has taken is already over 85%. He can get an 80% on the fifth test and still get an average of 85%. The sum of Charlie's 4 test scores is 345. In order to have a mean of 85 for all five scores, he needs a total of $(85)(5) = 425$. Then $425 - 345 = 80$, which is the minimum score needed on the 5th test.

39. D **(Standard Assessed: 10.D.1)**

A bar graph is the best way to display this information. Too much information is given for a circle graph (the percentages add up to over 100). There aren't two variables, so a line graph isn't the best choice, and there isn't enough information to plot on a scatter plot.

40. C **(Standard Assessed: 10.P.1)**

This pattern is the Fibonacci Numbers, where the two previous numbers are added together. The sequence of the first twelve numbers should be shown, that are: 2, 3, 5, 8, 13, 21, 34, 55, 89, 144, 233, 377.

41. **(Standard Assessed: 10.P.8)**

Sample answer:

a. $c = 100 + 10g$

b. $c = 100 + 10(112)$; $c = \$1,220$

c. $c = 12g$

d. Nora's Catering Service would charge \$1,220; Villa Mambo's Take-Out would charge \$1,344, so it is the more expensive of the two.

42. **(Standard Assessed: 10.D.2)**

Sample answer:

a.

mean: $0 + 0 + 1 + 2 + 4 + 4 + 4 + 6 + 7 + 8 = 36/10 = 3.6$
median: 4
mode: 4
range: $8 - 0 = 8$

b. actual scores: 80, 80, 81, 82, 84, 84, 84, 86, 87, 88

mean: 83.6
median: 84
mode: 84
range: $88 - 80 = 8$

Practice Test 2
SESSION 1

1 Ⓐ Ⓑ Ⓒ Ⓓ   **2** Ⓐ Ⓑ Ⓒ Ⓓ   **3** Ⓐ Ⓑ Ⓒ Ⓓ

4 Ⓐ Ⓑ Ⓒ Ⓓ   **5** Ⓐ Ⓑ Ⓒ Ⓓ   **6** Ⓐ Ⓑ Ⓒ Ⓓ

7 Ⓐ Ⓑ Ⓒ Ⓓ   **8** Ⓐ Ⓑ Ⓒ Ⓓ   **9** Ⓐ Ⓑ Ⓒ Ⓓ

10 Ⓐ Ⓑ Ⓒ Ⓓ   **11** Ⓐ Ⓑ Ⓒ Ⓓ   **12** Ⓐ Ⓑ Ⓒ Ⓓ

13 Ⓐ Ⓑ Ⓒ Ⓓ   **14** Ⓐ Ⓑ Ⓒ Ⓓ

15

16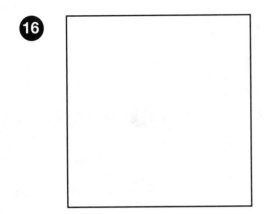

17 Ⓐ Ⓑ Ⓒ Ⓓ

18

19

20

21 _____

_____

_____

_____

_____

_____

_____

_____

_____

_____

_____

_____

_____

_____

_____

SESSION 2

22 Ⓐ Ⓑ Ⓒ Ⓓ   **23** Ⓐ Ⓑ Ⓒ Ⓓ   **24** Ⓐ Ⓑ Ⓒ Ⓓ

25 Ⓐ Ⓑ Ⓒ Ⓓ   **26** Ⓐ Ⓑ Ⓒ Ⓓ

27

28 Ⓐ Ⓑ Ⓒ Ⓓ   **29** Ⓐ Ⓑ Ⓒ Ⓓ   **30** Ⓐ Ⓑ Ⓒ Ⓓ

31

_____
_____
_____
_____
_____
_____
_____
_____
_____
_____

32 Ⓐ Ⓑ Ⓒ Ⓓ   **33** Ⓐ Ⓑ Ⓒ Ⓓ   **34** Ⓐ Ⓑ Ⓒ Ⓓ

35 Ⓐ Ⓑ Ⓒ Ⓓ   **36** Ⓐ Ⓑ Ⓒ Ⓓ   **37** Ⓐ Ⓑ Ⓒ Ⓓ

38 Ⓐ Ⓑ Ⓒ Ⓓ   **39** Ⓐ Ⓑ Ⓒ Ⓓ   **40** Ⓐ Ⓑ Ⓒ Ⓓ

41 _____
_____
_____
_____
_____

42